# Basic Math and Pre-Algebra

by Carolyn Wheater

ALPHA
A member of Penguin Group (USA) Inc.

**ALPHA BOOKS**

Published by Penguin Group (USA) Inc.

Penguin Group (USA) Inc., 375 Hudson Street, New York, New York 10014, USA • Penguin Group (Canada), 90 Eglinton Avenue East, Suite 700, Toronto, Ontario M4P 2Y3, Canada (a division of Pearson Penguin Canada Inc.) • Penguin Books Ltd., 80 Strand, London WC2R 0RL, England • Penguin Ireland, 25 St. Stephen's Green, Dublin 2, Ireland (a division of Penguin Books Ltd.) • Penguin Group (Australia), 250 Camberwell Road, Camberwell, Victoria 3124, Australia (a division of Pearson Australia Group Pty. Ltd.) • Penguin Books India Pvt. Ltd., 11 Community Centre, Panchsheel Park, New Delhi—110 017, India • Penguin Group (NZ), 67 Apollo Drive, Rosedale, North Shore, Auckland 1311, New Zealand (a division of Pearson New Zealand Ltd.) • Penguin Books (South Africa) (Pty.) Ltd., 24 Sturdee Avenue, Rosebank, Johannesburg 2196, South Africa • Penguin Books Ltd., Registered Offices: 80 Strand, London WC2R 0RL, England

IDIOT'S GUIDES and Design are trademarks of Penguin Group (USA) Inc.

International Standard Book Number: 978-1-61564-504-6
Library of Congress Catalog Card Number: 20148930954

16   15   14        8  7  6  5  4  3  2  1

Interpretation of the printing code: The rightmost number of the first series of numbers is the year of the book's printing; the rightmost number of the second series of numbers is the number of the book's printing. For example, a printing code of 14-1 shows that the first printing occurred in 2014.

*Printed in the United States of America*

**Note:** This publication contains the opinions and ideas of its author. It is intended to provide helpful and informative material on the subject matter covered. It is sold with the understanding that the author and publisher are not engaged in rendering professional services in the book. If the reader requires personal assistance or advice, a competent professional should be consulted. The author and publisher specifically disclaim any responsibility for any liability, loss, or risk, personal or otherwise, which is incurred as a consequence, directly or indirectly, of the use and application of any of the contents of this book.

Most Alpha books are available at special quantity discounts for bulk purchases for sales promotions, premiums, fundraising, or educational use. Special books, or book excerpts, can also be created to fit specific needs. For details, write: Special Markets, Alpha Books, 375 Hudson Street, New York, NY 10014.

**Publisher:** *Mike Sanders*
**Executive Managing Editor:** *Billy Fields*
**Executive Acquisitions Editor:** *Lori Cates Hand*
**Development Editor:** *Ann Barton*
**Senior Production Editor:** *Janette Lynn*

**Cover Designer:** *Laura Merriman*
**Book Designer:** *William Thomas*
**Indexer:** *Tonya Heard*
**Layout:** *Brian Massey*
**Proofreader:** *Gene Redding*

# Contents

## Appendixes

# Introduction

My job has always been teaching. Even when I wasn't officially working as a teacher, I was always explaining something to someone. Helping people understand new things was always what I ended up doing, whether it was running lunch hour calculus lessons for my senior classmates, explaining to my daughter how to solve systems of equations with matrices as we drove along a dark country road, or emailing explanations of linear programming or third grade multiplication to friends and family across the country. So it's not really a surprise that I'm writing this for you.

I don't know if I'm a "typical" teacher, but there are two ideas that have always guided my teaching. The first is that successful teaching and successful learning require that the teacher understand what the student doesn't understand. That doesn't just mean that the teacher is better educated. It means that the person doing the teaching actually sees why the other person finds an idea difficult or confusing. People tend to become teachers because they're good at a subject, but people who are good at a subject sometimes find it hard to see what's difficult and why. I've spent almost 40 years trying to understand, and I'm grateful to the hundreds of students who have taught me. I've tried to bring that understanding to this book.

The other guiding principle is that the teacher's job is to find another way to explain. And another, and another, and another, until one works. In my classrooms, that has led to silly stories about sheep, rules and formulas set to music, and quizzes that students giggle their way through. Whatever works, works, and language isn't just for language classes. How you tell the story can make all the difference for understanding it. I've tried to give you the benefit of what my students have taught me about the ways to explain math that work for them.

Part of the successful storytelling and the successful learning is creating a world your readers can imagine, visualize, and understand. This book is my attempt to take you into the world of numbers for a work-study tour. I hope you'll enjoy the trip.

## How This Book Is Organized

This book is presented in five sections.

In **Part 1, The World of Numbers,** you'll journey from the counting numbers, through the integers, and on to the rational numbers and the irrational numbers. You'll take a tour of the universe that mathematicians call the real numbers. This is no sightseeing tour. You'll work your way through the natural numbers, the integers, and the rational numbers, presented as both fractions and decimals. You'll practice all the arithmetic you need to know and explore different ways of writing numbers and the relationships among them.

In **Part 2, Into the Unknown,** you'll venture into the realm of variables and get acquainted with algebra. You'll solve equations and inequalities and graph them and begin to think about undoing arithmetic instead of doing it.

In **Part 3, The Shape of the World,** you'll take some basic ideas like measurement, congruence, proportion, and area and examine how they show up when you work with different types of geometric figures.

In **Part 4, The State of the World,** it's time to think about the chances and the risks and to report on the facts and figures that summarize what you've learned about the world.

**Part 5, Extra Practice,** is just what it sounds like. Building math skills is like learning to play an instrument: you have to do it, again and again, before you really get to be good at it. When you've traveled around the world, it's natural to want to go back and remember what you've seen. This is your chance.

## Extras

As you make your way through the world of numbers, you'll see some items set off in ways meant to catch your attention. Here's a summary of what you'll see.

> **CHECK POINT**
>
> As you take your world tour, you'll find that from time to time you need to pass through a Check Point. No passport required on our tour, but you will be asked to answer a few key questions to see if you're ready to move on. You'll find the answers for these Check Point questions in Appendix C.

> **DEFINITION**
>
> For a successful trip, it's a good idea to speak at least a little bit of the language of the area you're visiting. The Speak the Language sidebars throughout this book identify critical words and phrases that you'll want to know and use.

> **MATH TRAP**
>
> Ah, the unsuspecting tourist! It's so easy for someone who's just visiting to be fooled or to make embarrassing mistakes. Don't be that person. These sidebars serve as a caution and try to help you think and act like someone who calls the world of mathematics home.

**WORLDLY WISDOM**

There you are, you savvy sightseer, visiting new places and learning new things. Watch for these sidebars that point out bits of information and insight about the world of mathematics.

**MATH IN THE PAST**

Over the centuries that people have studied mathematics, their way of writing numbers, performing calculations, and organizing their thinking about math have grown and changed. Some of those ideas are still with us, some have faded away, and some have led to important discoveries. These sidebars will highlight some of these historical developments that connect to your current studies.

## Acknowledgments

It's always hard to know who to mention at this point in a book. You, my intended reader, may have no idea who these people are, and you may skip over this section because of that. Or you may read this and wonder if these folks are as strange as I am. The most important people may never see the book, and yet they should be mentioned.

So I'll begin by being forever grateful to E. Jones Wagner—Jonesy—who took a chance on an eager but very inexperienced young teacher. Jonesy showed me that different students learn in different ways and different teachers teach in different ways, and that to be a successful teacher, I had to find my own way. She helped me look past a lifetime of "shoulds" to what actually worked. She taught me, by her counsel, her example, her style, and yes, her eccentricity. Forty years later, I still think back to what I learned from Jonesy and, when faced with a problem, wonder what Jonesy would do. And to this day, if I see anything yellow or orange around the school house, I still want to return it to Jonesy's classroom.

My gratitude goes to Grace Freedson, of Grace Freedson's Publishing Network, who not only won't let me get lazy but also offers me projects, like this one, that are satisfying and challenging, and help me to grow as a teacher and as a person. My thanks also go to Lori Hand and Ann Barton for making this project an absolute delight, from start to finish, and for making my scribblings about math look good and make sense.

One of the things I tell my students is that it's normal, natural, even valuable to make mistakes. It's how we learn. Or more correctly, correcting our mistakes is how we learn. We all make mistakes. I certainly do, which is why there is someone who reads this math before you do. The Technical Reviewer's job is to read everything I've written about the math and make sure it's correct and clear. That job also includes checking all the problems and the answers and finding my mistakes. Yes, I made mistakes, and I am grateful to my Technical Reviewer for finding them and pointing them out to me. It makes a better book for you, and it teaches me about where errors might occur and how my brain works.

Finally, to my family—Laura Wheater, Betty and Tom Connolly, Frank and Elly Catapano—who patiently put up with my tendency to be just a tad obsessive when I'm working on a project, and to Barbara, Elise, and Pat, who keep me grounded and sort of sane, I send a giant thank you.

## Special Thanks to the Technical Reviewer

*Idiot's Guides: Basic Math and Pre-Algebra* was reviewed by an expert who double-checked the accuracy of what you'll learn here, to help us ensure this book gives you everything you need to know about basic math. Special thanks are extended to Steve Reiss.

## Trademarks

All terms mentioned in this book that are known to be or are suspected of being trademarks or service marks have been appropriately capitalized. Alpha Books and Penguin Group (USA) Inc. cannot attest to the accuracy of this information. Use of a term in this book should not be regarded as affecting the validity of any trademark or service mark.

# The World of Numbers

Welcome to the world of numbers! Any study of mathematics begins with numbers. Your sense of how much, how many, and how big or small is critical to the work you do in math, as well as to your understanding the environment in which you function.

The world of numbers is a wide world. In this part, you'll look at the areas of it we visit most often, but you won't have time to explore every corner of the world. Think of this as a tour to get acquainted with numbers. You'll learn to communicate in mathematical language and accomplish basic tasks. You'll learn the fundamental rules and relationships of our number system.

# Our Number System

When asked to think about the word "math," the first image most people are likely to have is one that involves numbers. This makes sense, because most of what we do in the name of math uses numbers in one way or another. Some would say that math is really about patterns, and that numbers and shapes are the vehicles, so arithmetic and geometry become two primary areas of mathematical thinking. There's a wider world to mathematics, but you have to start somewhere, and generally you start with numbers.

In this chapter, we'll take a look at the system of numbers we most commonly use. We'll explore how the system works and learn to identify the value of a digit based upon its position in the number. We'll examine how our system deals with fractions, or parts of a whole, and we'll explain some variations in the way numbers are written, techniques to avoid long strings of zeros, and a method of writing very large and very small numbers called scientific notation.

## In This Chapter

* The development of our number system
* Place value and how to use it
* How to express powers of ten using exponents
* Writing very large numbers in scientific notation

# The Counting Numbers

People have a tendency to think that our number system was always there and was always as it is now. On some level, that's true. The desire, and need, to count things dates to early history, but how people count and what people do with numbers have changed over the years. The need to count is so fundamental that the whole system is built on the numbers people use to count. The *counting numbers*, also called the *natural numbers*, are the numbers 1, 2, 3, 4, and so on. The counting numbers are an infinite set; that is, they go on forever.

You might notice that the counting numbers don't include 0. There's a simple reason for that. If you don't have anything, you don't need to count it. Zero isn't a counting number, but for reasons you'll see shortly, it's one that is used a lot. The set of numbers 0, 1, 2, 3, 4, and so on is called the *whole numbers*.

> **DEFINITION**
>
> The **counting numbers** are the set of numbers {1, 2, 3, 4, ...}. They are the numbers we use to count. The counting numbers are also called the **natural numbers**.
>
> The **whole numbers** are the set of numbers {0, 1, 2, 3, 4, ...} They are formed by adding a zero to the counting numbers.

Numbers didn't always look like they do now. At different times in history and in different places in the world, there were different symbols used to represent numbers. If you think for a moment, you can probably identify a way of writing numbers that is different from the one you use every day. Roman numerals are an ancient system still used in some situations, often to indicate the year. The year 2013 is MMXIII, and the year 1960 is MCMLX.

Roman numerals choose a symbol for certain important numbers. I is 1, V is 5, X stands for 10, L for 50, C for 100, D for 500 and M for 1,000. Other numbers are built by combining and repeating the symbols. The 2000 in 2013 is represented by the two Ms. Add to that an X for 10 and three Is and you have 2013. Position has some meaning. VI stands for 6 but IV stands for 4. Putting the I before the V takes one away, but putting it after adds one. Roman numerals obviously did some jobs well or you wouldn't still see them, but you can probably imagine that arithmetic could get very confusing.

> **MATH IN THE PAST**
>
> Ever wonder why the Romans chose those letters to stand for their numbers? They may not have started out as letters. One finger looks like an I. Hold up your hand to show five fingers and the outline of your hand makes a V. Two of those, connected at the points, look like an X and show ten.

The ancient Romans weren't the only culture to have their own number system. There were many, with different organizing principles. The system most commonly in use today originated with Arabic mathematicians and makes use of a positional, or *place value system*. In many ancient systems, each symbol had a fixed meaning, a set value, and you simply combined them. In a place value system, each position represents a value and the symbol you place in that position tells how many of that value are in the number.

**DEFINITION**

A **place value system** is a number system in which the value of a symbol depends on where it is placed in a string of symbols.

## The Decimal System

Our system is a positional, or place value system, based on the number 10, and so it's called a *decimal system*. Because it's based on 10, our system uses ten *digits*: 0, 1, 2, 3, 4, 5, 6, 7, 8, and 9. A single digit, like 7 or 4, tells you how many ones you have.

**DEFINITION**

A **digit** is a single symbol that tells how many. It's also a word that can refer to a finger (another way to show how many).

When you start to put together digits, the rightmost digit, in the ones place, tells you how many ones you have, the next digit to the left is in the tens place and tells how many tens, and the next to the left is how many hundreds. That place is called the hundreds place. The number 738 says you have 7 hundreds, 3 tens, and 8 ones, or seven hundred thirty-eight.

The number 392,187 uses six digits, and each digit has a place value. Three is in the hundred-thousands place, 9 in the ten-thousands place, and 2 in the thousands place. The last three digits show a 1 in the hundreds place, 8 in the tens place and 7 in the ones place.

### The Meaning of Digits in a Place Value System

| Place Name | Hundred-thousands | Ten-thousands | Thousands | Hundreds | Tens | Ones |
|---|---|---|---|---|---|---|
| Value | 100,000 | 10,000 | 1,000 | 100 | 10 | 1 |
| Digit | 3 | 9 | 2 | 1 | 8 | 7 |
| Worth | 300,000 | 90,000 | 2,000 | 100 | 80 | 7 |

📖 **DEFINITION**

A **decimal system** is a place value system in which each position in which a digit can be placed is worth ten times as much as the place to its right.

Each move to the left multiplies the value of a digit by another 10. The 4 in 46 represents 4 tens or forty, the 4 in 9,423 is 4 hundreds, and the 4 in 54,631 represents 4 thousands.

If you understand the value of each place, you should be able to tell the value of any digit as well as the number as a whole.

✏️ **CHECK POINT**

Complete each sentence correctly.

1. In the number 3,492, the 9 is worth _____.

2. In the number 45,923,881, the 5 is worth _____.

3. In the number 842,691, the 6 is worth _____.

4. In the number 7,835,142, the 3 is worth _____.

5. In the number 7,835,142, the 7 is worth _____.

When you read a number aloud, including an indication of place values helps to make sense of the number. Just reading the string of digits "three, eight, two, nine, four" tells you what the number looks like, but "thirty-eight thousand, two hundred ninety-four" gives you a better sense of what it's worth.

In ordinary language, the ones place doesn't say its name. If you see 7, you just say "seven," not "seven ones." The tens place has the most idiosyncratic system. If you see 10, you say "ten," but 11 is not "ten one." It's "eleven" and 12 is "twelve," but after that, you add "teen" to the ones digit. Sort of. You don't have "threeteen," but rather "thirteen." You do have "fourteen" but then "fifteen." The next few, "sixteen," "seventeen," "eighteen," and "nineteen" are predictable.

When the tens digit changes to a 2, you say "twenty" and 3 tens are "thirty," followed by "forty," "fifty," "sixty," "seventy," "eighty," and "ninety." Each group of tens has its own family name, but from twenty on, you're consistent about just tacking on the ones. So 83 is "eighty-three" and 47 is "forty-seven." And the hundreds? They just say their names.

Larger numbers are divided into groups of three digits, called periods. A period is a group of three digits in a large number. The ones, tens and hundreds form the ones period. The next three digits are the thousands period, then the millions, the billions, trillions, and on and on.

**WORLDLY WISDOM**

In the United States, you separate periods with commas. In other countries, like Italy, they're separated by periods, and in others, like Australia, by spaces.

You read each group of three digits as if it were a number on its own and then add the period name. The number 425 is "four hundred twenty-five," so if you had 425,000, you'd say "four hundred twenty-five thousand." The number 425,000,000 is "four hundred twenty-five million," and 425,425,425 is "four hundred twenty-five million, four hundred twenty-five thousand, four hundred twenty-five."

**CHECK POINT**

6. Write the number 79,038 in words.

7. Write the number 84,153,402 in words.

8. Write "eight hundred thirty-two thousand, six hundred nine" in numerals.

9. Write "fourteen thousand, two hundred ninety-one" in numerals.

10. Write "twenty-nine million, five hundred three thousand, seven hundred eighty-two" in numerals.

## Powers of Ten

Each place in a decimal system is ten times the size of its neighbor to the right and a tenth the size of its neighbor to the left. As you move through a number, there are a whole lot of tens being used. You can write out the names of the places in words: the hundredths place or the ten-thousands place. You can write their names using a 1 and zeros: the 100 place or the 10,000 place. The first method tells you what the number's name sounds like, and the other helps you have a sense of what the number will look like.

You can keep moving into larger and larger numbers, and the naming system keeps going with the same basic pattern. The problem is that those numbers, written in standard notation, take up lots of space and frankly, don't always communicate well. In standard notation, one hundred trillion is 100,000,000,000,000. Written that way, most of us just see lots of zeros, and it's hard to register how many and what they mean.

There's a shortcut for writing the names of the places called *powers of ten*. All of the places in our decimal system represent a value that's written with a 1 and some zeros. The number of zeros depends on the place. The ones place is just 1—no zero. The tens place is 10, a 1 and one zero. The hundreds place is 100, a 1 and two zeros. The thousands place has a value of 1,000 or a 1 and three zeros, and on it goes.

To write powers of ten in a more convenient form, you use *exponents*. These are small numbers that are written to the upper right of another number, called the base, and tell how many of that number to multiply together.

If you want to show $3 \times 3$, you can write $3^2$. In this case, 3 is the base number and 2 is the exponent. This notation tells you to use two 3s and multiply them. We'll look at exponents again in a later chapter, but for now we're going to take advantage of an interesting result of working with tens.

---

**DEFINITION**

The expression **power of ten** refers to a number formed by multiplying a number of 10s. The first power of ten is 10. The second power of ten is $10 \times 10$ or 100, and the third power of 10 is $10 \times 10 \times 10$ or 1,000.

An **exponent** is a small number written to the upper right of another number, called the base. The exponent tells how many of that number should be multiplied together. You can write the third power of 10 (10 is $10 \times 10 \times 10$) as $10^3$. In this case, 10 is the base number and 3 is the exponent.

---

When you multiply tens together, you just increase the number of zeros. $10 \times 10 = 100$, $100 \times 10 = 1000$. Each time you multiply by another ten, you add another zero. Look at a place value, count the number of zeros in the name, and put that exponent on a 10, and you have the power-of-ten form of that place value.

## Powers of 10

| Decimal Place | Value | Number of Zeros | Tens being multiplied | Power of Ten |
|---|---|---|---|---|
| ones | 1 | 0 | None | $10^0$ |
| tens | 10 | 1 | 10 | $10^1$ |
| hundreds | 100 | 2 | $10 \times 10$ | $10^2$ |
| thousands | 1,000 | 3 | $10 \times 10 \times 10$ | $10^3$ |

Using this system, a million, which you write as 1,000,000 in standard notation, has 6 zeros after the 1, so it would be $10^6$. One hundred trillion is 100,000,000,000,000 or a 1 followed by 14 zeros. You can write one hundred trillion as $10^{14}$, which is a lot shorter.

CHECK POINT

11. Write 10,000 as a power of ten.

12. Write 100,000,000,000 as a power of ten.

13. Write $10^7$ in standard notation.

14. Write $10^{12}$ in standard notation.

15. Write $10^5$ in standard notation.

# Scientific Notation

Suppose you needed to talk about the distance from Earth to Mars (which keeps changing because both planets are moving, but you can give an approximate distance). You can say that Earth and Mars are at least 34,796,800 miles apart and probably not more than 249,169,848 miles apart, so on average, about 86,991,966.9 miles. If you read that last sentence and quickly lost track of what the numbers were and replaced their names with a mental "oh, big number," you're not alone.

Whether they're written as a string of digits like 34,796,800 or in words like two hundred forty-nine million, one hundred sixty-nine thousand, eight hundred forty-eight, our brains have trouble really making sense of numbers that large. (Whether you think the numbers or the words are easier to understand is a personal matter. Our brains are not all the same.) Scientists and others who work with very large or very small numbers on a regular basis have a method for writing such numbers, called scientific notation.

*Scientific notation* is a system of expressing numbers as a number between one and ten, times a power of ten. The first number is always at least 1 and less than 10. Ten and any number bigger than ten can be written as a smaller number times a power of ten.

Let's look at that with a few smallish numbers first. A single digit number like 8 would be $8 \times 10^0$. Ten to the zero power is 1, so $8 \times 10^0$ is $8 \times 1$ or 8. The number 20 would be $2 \times 10^1$. $10^1$ is 10, so $2 \times 10^1$ is $2 \times 10$, or 20. For a larger number like 6,000,000 you would think of it as $6 \times 1,000,000$, or $6 \times 10^6$.

DEFINITION

**Scientific notation** is a method for expressing very large or very small numbers as the product of a number between 1 and 10 and a power of 10.

To write a large number in scientific notation, copy the digits and place a decimal point after the first digit. This creates the number between 1 and 10. Count the number of places between where you just put the decimal point and where it actually belongs. This is the exponent on the ten. Once you write the number as a number between 1 and 10 times a power of 10, you can drop any trailing zeros, zeros at the end of the number.

Here's how to write 83,900 in scientific notation:

1. Write the digits without a comma.
   83900

2. Insert a decimal point after the first digit.
   8.3900

3. Count the places from where the decimal is now to where it was originally.
   8.3900
   $\overrightarrow{4\ places}$

4. Write as a number between 1 and 10 multiplied by a power of 10.
   $8.3900 \times 10^4$

5. Drop trailing zeros.
   $8.39 \times 10^4$

The number 83,900 can be written as $8.39 \times 10^4$.

To change a number that is written in scientific notation to standard notation, copy the digits of the number between 1 and 10 and move the decimal point to the right as many places as the exponent on the 10. You can add zeros if you run out of digits. The number $3.817 \times 10^8$ becomes 3.$\underset{8\ places}{\underline{81700000}}$ or 381,700,000.

---

**CHECK POINT**

16. Write 59,400 in scientific notation.

17. Write 23,000,000 in scientific notation.

18. Write $5.8 \times 10^9$ in standard notation.

19. Write $2.492 \times 10^{15}$ in standard notation.

20. Which is bigger: $1.2 \times 10^{23}$ or $9.8 \times 10^{22}$?

# Rounding

When dealing with large quantities, sometimes you don't need to use exact numbers. If you want to talk about a number being "about" or "approximately," you want to round the number. For example, the number 6,492,391 is closer to 6 million than to 7 million, but closer to 6,500,000 than to 6,400,00. Rounding is a process of finding a number with the desired number of *significant digits* that is closest to the actual number.

> **DEFINITION**
>
> The significant digits of a number are the nonzero digits and any zeros that serve to tell you the precision of the measurement or the digit to which the number was rounded.

When you round a number, you place it between two other numbers and decide to which it is closer. To round 48,371 to the nearest ten-thousand, you need to decide if it's closer to 40,000 or to 50,000. Any number from 40,001 up to 44,999 would be closer to 40,000, but numbers from 45,001 to 49,999 are closer to 50,000. The general agreement is that 45,000, right in the middle, will round to 50,000.

Because that middle number is the dividing line between the numbers that round down and those that round up, the digit after the last significant digit will tell you which way to round. If you want to round 48,371 to the nearest thousand, look to the hundreds place. The digit in that place is 3, so round down to 48,000. If you want to round it to the nearest hundred, the 7 in the tens place tells you to round up to 48,400.

To round a number:

1. Decide how many significant digits you want to keep.

2. Look at the next digit to the right.

3. If that digit is less than 5, keep the significant digits as they are and change the rest of the digits to zeros.

4. If that digit is 5 or more, increase the last significant digit by one and change the following digits to zeros.

Don't worry if you start to round up and feel like you've started a chain reaction. If you round 99,999 to the nearest hundred, you're placing 99,999 between 99,900 and the number 100 higher, which is 100,000. You see the 9 in the tens place and know you need to round up. That means you need to change the 9 in the hundreds place to a 10, and that doesn't fit in one digit. That extra digit is carried over to the thousands place, which makes that a 10, and that carries over to the ten-thousands place. Take a moment to think about what numbers you're choosing between, and you'll know you're in the right place.

CHECK POINT

Round each number to the specified place.

21. 942 to the nearest hundred

22. 29,348 to the nearest ten-thousand

23. 1,725,854 to the nearest hundred-thousand

24. 1,725,854 to the nearest thousand

25. 1,725,854 to the nearest million

## The Least You Need to Know

- Our number system is a place value system based on powers of ten.
- As you move to the left, the value of each place is multiplied by 10.
- An exponent is a small number written to the upper right of a base number. The exponent tells you how many of the base number to multiply together.
- Scientific notation is a system of writing large numbers as a number between 1 and 10 multiplied by a power of 10.
- Round a number to a certain place by looking at the next place, rounding up if the next digit is 5 or more and down it's 4 of less.

# CHAPTER 2

# Arithmetic

The last chapter focused on the world of numbers and how to express those numbers in words and symbols. The next step is investigating how to work with numbers. In other words, it's time to look at arithmetic.

The two fundamental operations of arithmetic are addition and multiplication (and multiplication is actually a shortcut for repeated addition). Subtraction and division are usually included in the basics of arithmetic, but you'll see that these are really operations that reverse on addition and multiplication. We'll look at the basics of skillful arithmetic and introduce some strategies that may make the work easier.

## Addition and Subtraction

The counting numbers came to be because people needed to count. Soon thereafter, people started putting together and taking apart the things, or groups of things, they had counted. If you have 3 fish and your best buddy has 5 fish, you have a pretty satisfying meal (unless you invite 20 friends, but that's division and that's later). You could put all the fish in a pile and count them again, but soon you get the notion of 3 + 5 = 8. And if you only eat 4 fish, again, you could count the fish that are left, but before too long, you grasp subtraction: $8 - 4 = 4$. Addition and subtraction are just ideas of putting together and taking away that people developed to avoid having to keep recounting.

## Addition

If addition is just a substitute for counting, why do you need it? The need is speed, especially when the numbers get large. Counting 3 fish and 5 fish to find out you have 8 fish is fine, but counting 3,000 fish and 5,000 fish would not be practical. When the *addends,* the numbers you're adding, get larger, you need a system to find the *sum,* or total, quickly. The place value system allows you to add large numbers in a reasonable amount of time.

> **DEFINITION**
>
> A **sum** is the result of addition. The numbers that are added are called **addends**. In the equation 5 + 10 = 15, 5 and 10 are addends, and 15 is the sum.

It's important, of course, to learn basic addition facts, but the facts in the table and an understanding of our place value system will get you through most addition problems.

### Basic Addition Facts

| + | 1 | 2 | 3 | 4 | 5 | 6 | 7 | 8 | 9 |
|---|---|---|---|---|---|---|---|---|---|
| **1** | 2 | 3 | 4 | 5 | 6 | 7 | 8 | 9 | 10 |
| **2** | 3 | 4 | 5 | 6 | 7 | 8 | 9 | 10 | 11 |
| **3** | 4 | 5 | 6 | 7 | 8 | 9 | 10 | 11 | 12 |
| **4** | 5 | 6 | 7 | 8 | 9 | 10 | 11 | 12 | 13 |
| **5** | 6 | 7 | 8 | 9 | 10 | 11 | 12 | 13 | 14 |
| **6** | 7 | 8 | 9 | 10 | 11 | 12 | 13 | 14 | 15 |
| **7** | 8 | 9 | 10 | 11 | 12 | 13 | 14 | 15 | 16 |
| **8** | 9 | 10 | 11 | 12 | 13 | 14 | 15 | 16 | 17 |
| **9** | 10 | 11 | 12 | 13 | 14 | 15 | 16 | 17 | 18 |

The left column and top row show the digits from 1 to 9. The box where a row and column meet contains the sum of those digits.

When you add 3 + 5, you get 8, a single digit. That could be adding 3 ones and 5 ones to get 8 ones, but 3 thousands + 5 thousands make 8 thousands, so the same addition could be done for larger addends. It's just about writing the digits in the proper places.

When you add numbers with more than one digit, stack them one under the other, with the decimal points aligned (even if the decimal points are unseen.) To add 43,502 and 12,381, you write them like this:

$$+\ \frac{\begin{array}{r}43,502\\12,381\end{array}}{}$$

This puts the ones under the ones, the tens under the tens, the hundreds under the hundreds, and so on. You just have to add the digits in each place, starting on the right with the ones place:

2 ones + 1 one equals 3 ones

0 tens + 8 tens equals 8 tens

5 hundreds + 3 hundreds equals 8 hundreds

3 thousands + 2 thousands equals 5 thousands

4 ten-thousands + 1 ten-thousand equals 5 ten-thousands

$$+\ \frac{\begin{array}{r}43,502\\12,381\end{array}}{55,883}$$

It's traditional to start from the right, from the lowest place value, and work up. There are problems, like this one, which could be done left to right, but in the next example, you'll see why right to left is the better choice. The key is that in the previous example, each time you added two digits, you got a single digit result. That's not always the case. Let's change just one digit in that problem. Change 43,502 to 43,572.

$$+\ \frac{\begin{array}{r}43,572\\12,381\end{array}}{}$$

Now when you add the ones digits you get a single digit, 3, which goes in the ones place of the answer, but when you add the 7 tens to the 8 tens, you get 15 tens, and there's no way to squeeze that two-digit 15 into the one space for the tens digit. You have to regroup, or as it's commonly called, you have to carry.

Our 15 tens can be broken up into one group of 10 tens and one group of 5 tens. The group of 10 tens makes 1 hundred. You're going to pass that 1 hundred over to the hundreds place, to the left, and just put the 5, for the remaining 5 tens, in the tens digit place. You put the 5 in the tens place of the answer, and place a small 1 above the hundreds column to remind yourself that you've passed 1 hundred along. In common language, you put down the 5 and carry the 1.

$$
\begin{array}{r}
\overset{1}{4}3,572 \\
+\ 12,381 \\
\hline
53
\end{array}
$$

When you add the hundreds digits, you add on that extra 1. Five hundreds + 3 hundreds + the extra 1 hundred from regrouping = 9 hundreds.

$$
\begin{array}{r}
\overset{1}{4}3,572 \\
+\ 12,381 \\
\hline
953
\end{array}
$$

As you finish the addition in the other columns, you'll find that each gives you just one digit, so no further regrouping or carrying is needed.

$$
\begin{array}{r}
\overset{1}{4}3,572 \\
+\ 12,381 \\
\hline
55,953
\end{array}
$$

A problem may not need regrouping at all, like our first example, or just once, like this example, or many times, or even every time. Say you want to add 9,999 and 3,457.

In the ones place, 9 + 7 = 16. Put down the 6 and carry the 1. In the tens, 9 + 5 + 1 you carried = 15. Put down the 5 and carry the 1. In the hundreds, 9 + 4 + 1 you carried = 14. Put down the 4 and carry the 1. In the thousands, 9 + 3 + 1 you carried = 13.

$$
\begin{array}{r}
\overset{1\ 1\ 1}{9},999 \\
+\ 3,457 \\
\hline
13,456
\end{array}
$$

---

✏️ **CHECK POINT**

Find each sum.

1. 48 + 86
2. 97 + 125
3. 638 + 842
4. 1,458 + 2,993
5. 12,477 + 8,394

## Adding Longer Columns

Addition is officially a binary operation. That means that it works with two numbers at a time. In fact, all four arithmetic operations (addition, subtraction, multiplication, and division) are binary operations, but for addition (and multiplication) it is possible to chain a series of operations together. When you add $4 + 9 + 7 + 5$, you're actually only adding two numbers at a time. You add $4 + 9$ to get 13, then that $13 + 7$ to get 20, and then $20 + 5$ to get 25. The same kind of chaining works with larger numbers, as long as you remember to stack the numbers so that decimal points, and therefore place values, are lined up.

Suppose you wanted to add $59,201 + 18,492 + 81,002 + 6,478$. First, stack them with the (invisible) decimal points aligned.

$$
\begin{array}{r}
59{,}201 \\
18{,}492 \\
+\ 81{,}002 \\
6{,}478 \\
\hline
\end{array}
$$

Add each column, regrouping and carrying over to the next place if you need to. Chain the addition in each column. In the ones column, you're adding $1 + 2 + 2 + 8$, so $1 + 2$ is 3, $3 + 2$ is 5, and $5 + 8$ is 13. Put the 3 in the ones column of the answer and carry the 1 ten to the tens column.

$$
\begin{array}{r}
59{,}2\overset{1}{0}1 \\
18{,}492 \\
+\ 81{,}002 \\
6{,}478 \\
\hline
3
\end{array}
$$

Add the tens column. $0 + 9 + 0 + 7 +$ the 1 you carried is 17, so put down the 7 and carry the 1.

$$
\begin{array}{r}
59{,}\overset{1\,1}{2}01 \\
18{,}492 \\
+\ 81{,}002 \\
6{,}478 \\
\hline
73
\end{array}
$$

Keep the chain of additions going, column by column, and don't be alarmed when the thousands column adds to 25. Just put down the 5 and carry a 2 over to the ten-thousands column.

$$
\begin{array}{r}
{\scriptstyle 2\,1\ \ 1\,1} \\
59,201 \\
18,492 \\
+\ \ 81,002 \\
6,478 \\
\hline
165,173
\end{array}
$$

There are a lot of little steps in an addition like this, and you may wonder why some people are so quick at it while others take more time. The answer probably has to do with the order in which those speedy folks do the job.

You can only add two numbers at a time, but you don't have to add them in the order they're given to you. There are two properties of addition that help to simplify and speed up your work. They're called the *commutative property* and the *associative property*.

The commutative property tells us that order doesn't matter when adding. It doesn't take long to realize that $4 + 7$ is the same as $7 + 4$, and the same is true for any two numbers you add. If you add $3,849,375 + 43,991$, you'll probably put the $3,849,375$ on top and the $43,991$ under that, because that's the way it was given to you, but you'd get exactly the same answer if you did it with the $43,991$ on the top.

It's important to remember that not every operation has this property. If you have $1,000 in the bank and withdraw $100, no one at the bank will blink, but if you have $100 and try to withdraw $1,000, there's likely to be an unpleasant reaction. Withdrawing money from your bank account is a subtraction, and subtraction isn't commutative.

> **DEFINITION**
>
> The **commutative property** is a property of addition or multiplication that says that reversing the order of the two numbers will not change the result.

The other useful property of addition is the *associative property*. This property tells us that when adding three or more numbers, you can group the addends in any combination without changing the outcome.

For example, when adding $8 + 5 + 5 + 2$, you might decide that adding those two 5s first is easier than going left to right. The associative property says you can do that without changing the result. You can take $8 + 5 + 5 + 2$, add the two 5s first, and make the problem $8 + 10 + 2$. That will give you 20, which is the same answer you would have gotten going left to right.

**DEFINITION**

The **associative property** is a property of addition or multiplication that says that when you must add or multiply more than two numbers, you may group them in different ways without changing the result.

If a problem has lots of numbers, you'll sometimes see parentheses around some of the numbers. This is a way of saying "do this part first." It might have $3 + (7 + 4) + 9$ to tell you to add the 7 and the 4 first. That can be helpful, and sometimes absolutely necessary, but the associative property says that if the problem is all addition or all multiplication, you can move those parentheses. It's telling you that you'll get the same answer no matter which part you do first, as long as the problem has just one operation, addition or multiplication. $3 + (7 + 4) + 9 = (3 + 7) + 4 + 9$. You can regroup.

When you let yourself use both the commutative and the associative properties, you realize that as long as addition is the only thing going on, you can tackle those numbers in any order. That's one step toward speedier work. If you can add any two numbers at a time, do the simple ones first. Adding $8 + 5 + 5 + 2$ is easier if you rearrange it to $5 + 5 + 8 + 2$. Add the 5s to get 10, add the $8 + 2$ to get another 10, and the two 10s give you 20.

**MATH TRAP**

Like the commutative property, the associative property is for addition (or multiplication), not for subtraction (or division) and not for combinations of operations. $(8 + 7) + 3$ is the same as $8 + (7 + 3)$, but $(8 + 7) \times 3$ is not the same as $8 + (7 \times 3)$.

Grouping the numbers in this way is easier because 5 and 5 are compatible numbers, and so are 8 and 2. Compatible numbers are pairs of numbers that add to ten. Why ten? Because our decimal system is based on tens. You might have your own version of compatible numbers, which add to something else.

For example, if you were a shepherd and had to keep track of a flock of seven sheep, you'd spend a lot of your time counting to seven to be sure you had them all. You'd quickly get to know that if you saw 5 of them in the pasture and 2 on the hill, all was well. If 3 were by the stream and 4 under the tree, you were good. You'd know all the pairs of numbers that added to 7: $1 + 6$, $2 + 5$, and $3 + 4$. For general addition purposes, however, compatible numbers are pairs that add to 10: $1 + 9$, $2 + 8$, $3 + 7$, $4 + 6$, and $5 + 5$.

Looking for compatible numbers, and taking advantage of the associative and commutative properties, can speed up addition. Adding $4 + 9 + 5 + 7 + 8 + 9 + 2 + 3 + 6 + 5 + 1$ just as written could be tedious. But if you look for compatible numbers and rearrange, it's not too bad.

$$\underline{4} + \underline{\underline{9}} + \overset{\frown}{5} + \overset{=}{7} + \overset{-}{8} + 9 + \overset{=}{2} + \overset{-}{3} + \underline{6} + \overset{\frown}{5} + \underline{\underline{1}}$$

$$= \underline{4 + 6} + \underline{\underline{9 + 1}} + \overset{\frown}{5 + 5} + \overline{\overline{7 + 3}} + \overline{8 + 2} + 9$$

$$= 10 + 10 + 10 + 10 + 10 + 9$$

$$= 59$$

If you use these tactics on each place value column, even long addition problems can go quickly. Take a look at the following addition problem.

$$
\begin{array}{r}
47,732 \\
18,481 \\
6,809 \\
+\phantom{0}14,678 \\
81,395 \\
\hline
\end{array}
$$

In the ones column, you can find $2 + 8$ and $1 + 9$ and another 5 to make 25. Put down the 5 and carry the 2. In the tens, there's a $3 + 7$ and an $8 + 2$ (that you carried) and another 9 for 29. Put down the 9 and carry the 2. In the hundreds, you'll find $7 + 3$, $4 + 6$ and $8 + 2$ for a total of 30. Put down the 0 and carry 3. There are two pairs that make 10 in the thousands, plus another 9, and the ten-thousands will add to 16.

$$
\begin{array}{r}
{\scriptstyle 2\,3\ \ 2\,2} \\
47,732 \\
18,481 \\
6,809 \\
+\phantom{0}14,678 \\
81,395 \\
\hline
169,095 \\
\end{array}
$$

**CHECK POINT**

Use compatible numbers to help you complete each addition problem.

6. $18 + 32 + 97$

7. $91 + 74 + 139$

8. $158 + 482 + 327 + 53$

9. $71,864 + 34,745 + 9,326$

10. $9,865 + 7,671 + 8,328 + 1,245 + 3,439$

# Subtraction

Subtraction is often thought of as a separate operation, but it's really a sort of backward addition. It's how you answer the question, "What do you add to A if you want to have B?" What do you add to 13 to get 20? Well, 20 − 13 is 7, which means 13 + 7 is 20. In this example, 7 is the *difference* between 20 and 13.

Subtraction and addition are *inverse operations*. Remember that shepherd who knows all the pairs that add to 7? If he looks up and only sees 5 sheep, he needs to know how many sheep are missing. His question can be phrased in addition terms as 5 + how many = 7? Or you can write it as a subtraction problem: 7 − 5 = how many? Either way you think about it, compatible numbers will be helpful with subtraction as well as addition.

> **DEFINITION**
>
> The result of a subtraction problem is called a **difference**. Officially, the number you start with is the **minuend**, and the number you take away is the **subtrahend**, but you don't hear many people use that language. In the equation 9 - 2 = 7, 9 is the minuend, 2 is the subtrahend, and 7 is the difference.
>
> An **inverse operation** is one that reverses the work of another. Putting on your jacket and taking off your jacket are inverse operations. Subtraction is the inverse of addition.

The "take-away" image of subtraction thinks of the problem as "if you have 12 cookies, and I take away 4, how many cookies are left?" That works fine for small numbers, and when you're working with large numbers, you can apply it one place value column at a time. In the following subtraction problem, you can work right to left:

$$\begin{array}{r} 6,987 \\ -\ 2,143 \\ \hline 4,844 \end{array}$$

7 ones take away 3 ones equals 4 ones

8 tens take away 4 tens equals 4 tens

9 hundreds take away 1 hundred equals 8 hundreds

6 thousands take away 2 thousands equals 4 thousands

Things get a little more complicated when you try a subtraction like 418 − 293. It starts out fine in the ones place: 8 − 3 = 5. But when you try to subtract the tens column, you have 1 − 9, and how can you take 9 away from 1? This is when you need to remember—and undo—the regrouping you did in addition.

In the process of addition, when the total of one column was more than one digit, too big to fit in that place, you carried some of it over to the next place. So when you're subtracting —going back—and you bump into a column that looks impossible, you're going to look to the next place up and take back, or borrow, so that the subtraction becomes possible.

In our example of 418 − 293, you'll look to the hundreds place of 418, where there are 4 hundreds, and you'll borrow 1 hundred. You'll cross out the 4 and make it a 3, so you don't forget that you borrowed 1 hundred, and you'll take that 1 hundred and change it back into 10 tens. You'll add those 10 tens to the 1 ten that was already in the tens place, and you'll have 3 hundreds, 11 tens, and 8 ones. Then you can take away 293, or 2 hundreds, 9 tens, and 3 ones. So 8 ones − 3 ones = 5 ones, 11 tens − 9 tens = 2 tens and 3 hundreds − 2 hundreds = 1 hundred. Here's how it would look:

$$
\begin{array}{r}
\overset{3}{\cancel{4}}\,\overset{11}{\cancel{1}}\,8 \\
-\ 2\ 9\ 3 \\
\hline
1\ 2\ 5
\end{array}
$$

Borrowing isn't always necessary, as you saw in the earlier example. When it is, you'll find it's wise to mark the ungrouping you've done and not try to juggle it all in your head.

---

### CHECK POINT

Complete each subtraction problem.

11. 596 − 312

12. 874 − 598

13. 1,058 − 897

14. 5,403 − 3,781

15. 14,672 − 5,839

---

Thinking of subtraction as adding back instead of taking away can be helpful for mental math. If you buy something that costs $5.98 and give the cashier a $10 bill, how much change should you get? Instead of doing all the borrowing and regrouping that's necessary to subtract 10.00 − 5.98, start with $5.98 and think about what you'd need to add to get to $10. You'd need 2 pennies, or $0.02, to make $6, and then another $4 to make $10. So your change should be $4.02.

You can use the add-back method of subtraction whenever it seems convenient, even if you're not making change. To subtract 5,250 − 3,825, you can start with 3,875 and think:

Adding 5 will make 3,830

Adding 20 will make 3,850

Adding 400 will make 4,250

Adding 1,000 will make 5,250

So 5,250 − 3,825 = 1,425.

Subtract by adding back.

16.  100 − 62

17.  250 − 183

18.  500 − 29

19.  400 − 285

20.  850 − 319

# Multiplication and Division

There's a lot of talk about the four operations of arithmetic: addition, subtraction, multiplication, and division. In a way, they all boil down to addition. As we've seen, subtraction is the inverse, or opposite, of addition. Multiplication is actually just a shortcut for adding the same number several times, and division is the inverse of multiplication. While each operation has its place, it's good to remember that they're all connected.

## Multiplication

Multiplication originated as a shorter way to express repeated addition. Suppose you pay $40 every month for your phone. How much is that per year? You could say that you pay $40 in each of the 12 months of the year and write an addition problem 40 + 40 + 40 + 40 + 40 + 40 + 40 + 40 + 40 + 40 + 40 + 40. It's long, but it would do the job. Multiplication lets you say the same thing as 40 × 12. (Or 12 × 40, thanks to the commutative property.) In this example, 40 and 12 are both called factors. When you multiply 40 by 12, you get 480, which is called the product.

DEFINITION

Each number in a multiplication problem is a **factor**. The result of the multiplication is the **product**. In the equation 5 × 3 = 15, 5 and 3 are factors, and 15 is the product.

Although it's nice to have a shorter way to write the problem, multiplication isn't much use to us unless it also gives us a simpler way to do the problem. And it will, but you need to do the memory work to learn the basic multiplication facts, or what most people call the times tables.

In the following chart, each column is one table or family of facts. The first column is the ones table, or what you'd get if you added one 1, two 1s, three 1s, and so on. The last column is the nines table. One 9 is 9, two 9s are 18, three 9s are 27, and on down to nine 9s are 81. (You'll see some tables that include 10 and sometimes even larger numbers, and the more tables you can learn, the faster at multiplication you'll be.)

## Basic Multiplication Facts

| × | 1 | 2 | 3 | 4 | 5 | 6 | 7 | 8 | 9 |
|---|---|---|---|---|---|---|---|---|---|
| 1 | 1 | 2 | 3 | 4 | 5 | 6 | 7 | 8 | 9 |
| 2 | 2 | 4 | 6 | 8 | 10 | 12 | 14 | 16 | 18 |
| 3 | 3 | 6 | 9 | 12 | 15 | 18 | 21 | 24 | 27 |
| 4 | 4 | 8 | 12 | 16 | 20 | 24 | 28 | 32 | 36 |
| 5 | 5 | 10 | 15 | 20 | 25 | 30 | 35 | 40 | 45 |
| 6 | 6 | 12 | 18 | 24 | 30 | 36 | 42 | 48 | 54 |
| 7 | 7 | 14 | 21 | 28 | 35 | 42 | 49 | 56 | 63 |
| 8 | 8 | 16 | 24 | 32 | 40 | 48 | 56 | 64 | 72 |
| 9 | 9 | 18 | 27 | 36 | 45 | 54 | 63 | 72 | 81 |

The first column and top row show the single digit numbers. Each cell shows the product of the digits that start its row and column.

Once you've memorized the multiplication facts, you can take advantage of what you know about our place value system to handle larger numbers. Let's start with something not too large. Let's multiply 132 × 3.

You know that 132 means 1 hundred and 3 tens and 2 ones, and you know that multiplying by 3 is the same as adding 132 + 132 + 132. Instead of all that adding, you can multiply each digit of 132 by 3, using the 3 times table, and you get 396 or 3 hundreds, 9 tens, and 6 ones. Each digit, each place, gets multiplied by 3. 132 × 3 = 396.

Now let's multiply 594 × 2. You have 5 hundreds, 9 tens, and 4 ones, and you want to multiply by 2. That should give us 10 hundreds, 18 tens, and 8 ones, but that means that you again face that problem of having a two-digit answer and only a one-digit place to put it. Just as you did with some of our addition, you're going to need to carry.

Once again, you want to tackle that work from right to left. So, in this example, you start with 2 × 4 = 8 and that's a single digit, so it can go in the ones place of the answer. Next, you multiply 2 × 9 = 18. That's 18 tens. You break that into 1 group of 10 tens and another 8 tens, or 1 hundred and 8 tens. The 8 can go in the tens place of your answer, but you'll hold on to the 1 hundred for a minute. Put a little 1 over the 5 to remind yourself that you have that 1 hundred waiting. One more multiplication, this time 2 × 5 = 10 hundreds, and then you'll add on the

1 hundred that's been waiting and you'll have 11 hundreds. Of course, 11 hundreds form 1 group of 10 hundreds plus 1 more hundred, or 1 thousand and 1 hundred. Our multiplication ends up looking like this.

$$594 \Rightarrow \overset{1}{594} \Rightarrow \overset{1}{594}$$
$$\underline{\times\ 2} \qquad \underline{\times\ 2} \qquad \underline{\times\ 2}$$
$$8 \qquad\ 88 \qquad 1188$$

When all the multiplying and regrouping is done, $594 \times 2 = 1{,}188$.

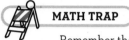 **MATH TRAP**

Remember that any carrying you do in multiplication happens after you do a multiplication. The digit you carry is part of the result of the multiplication. It has already been through the multiplication process, so make sure you wait and add it on after the next multiplication is done. Don't let it get into the multiplication again.

So our plan for multiplication of a larger number by a single digit is to start from the right, multiply each digit in the larger number by the one-digit multiplier, and carry when the result of a multiplication is more than one digit. That will work nicely when you want to multiply a larger number by one digit, but what if both of the numbers have more than one digit? What if you need to multiply 594 by 32 (instead of by just 2)?

For a problem like $594 \times 32$, you don't have to invent a new method, but you do have to adapt the method a little. You'll still start from the right and multiply each digit of 594 by 2, carrying when you need to. Then you'll multiply each digit of 594 by 3, again starting from the right. But here's the catch: that 3 you're multiplying by is 3 tens, not 3 ones. You have to work that change in place value into your multiplication.

The problem is fairly easy to solve. Just look at the first time you multiply by the 3 tens: $3 \times 4$. It's really 3 tens times 4, and that should give you not 12, but 12 tens or 120. Multiplying by 3 tens instead of just 3 simply adds a zero.

So here's how you'll tackle the problem. First you'll multiply 594 by the 2, just as you did before.

$$594$$
$$\underline{\times 32}$$
$$1188$$

Next, because you know that our next multiplication is by 3 tens and will add a zero, you'll put a zero under the ones digits. Then you'll multiply each digit of 594 by 3, carrying when you need to, and placing those digits on the line with that zero. It looks like this:

```
    2 1
  594
 × 32
 ─────
 1188
17820
```

You start the second line of multiplication with the 0 on the right end, then multiply 594 × 3 to get 1782. Finally, you'll add the two lines of results to get your final product.

```
    2 1
  594
 × 32
 ─────
 1 1
 1188
17820
─────
19008
```

When you're all done, you know that 594 × 32 = 19,008.

Notice that you've multiplied 594 by 2 and 594 by 30, then added those two products together.

The idea of adding that zero because you're multiplying by a digit in the tens place can be extended to other places. If you're going to multiply by a digit in the hundreds place, you'll put two zeros at the right end of that line. If the digit you're multiplying by is in the thousands place, you'll start the line with three zeros.

Here's a larger problem with very simple multiplication, so that you can see how the plan works.

```
    11,111
  ×   345
  ────────
    1 1
    55,555
   1
   444,440
 3,333,300
 ──────────
 3,833,295
```

You can see that the zeros shift each line of multiplication over one place, and that's because each new multiplication is by a digit that's worth ten times as much. The hardest part is all the adding, but when you're done 11,111 × 345 = 3,833,295.

**WORLDLY WISDOM**

There's a method of multiplication that never requires you to multiply more than one digit by one digit and takes care of carrying automatically. It's called lattice multiplication. You make a grid of boxes with diagonal lines from upper right to lower left. Factors are written across the top and down the right side, one digit per box. Each multiplication of one digit by one digit goes in a box, with the tens digit above the diagonal and the ones below. Add along the diagonals, and read the product down the left side and across the bottom.

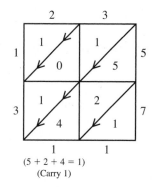

(5 + 2 + 4 = 1)
(Carry 1)

Are you ready for some multi-digit multiplication? Remember your basic facts, and don't forget to add zeros to keep the place value columns aligned.

**CHECK POINT**

Complete each multiplication problem.

21.  462 × 53

22.  833 × 172

23.  1,005 × 53

24.  1,841 × 947

25.  2,864 × 563

# Division

Division is the inverse, or opposite, of multiplication. If multiplication answers questions like "If I have 3 boxes of cookies and each box has 12 cookies, how many cookies do I have?" then division is for questions like "If I have 36 cookies and I'm going to put them in 3 boxes, how many cookies go in each box?" The division problem "36 ÷ 3 = what number?" is equivalent to "3 × what number = 36?"

 **DEFINITION**

The result of a division is called a **quotient**. The number you divide by is the **divisor**, and the number you're dividing is called the **dividend**. Dividend ÷ divisor = quotient. In the equation 12 ÷ 3 = 4, 12 is the dividend, 3 is the divisor, and 4 is the quotient.

Just as knowing your addition facts helps with subtraction, knowing your multiplication facts, or times tables, will help with division. And just as you used the place value system to deal with larger numbers in other operations, a process called long division will use a similar strategy.

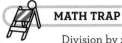 **MATH TRAP**

Division by zero is impossible. If you divide 12 by 4, you're asking 4 × what number = 12. If you try to divide 12 by 0, you're asking 0 × what number = 12, and the answer is there isn't one. Zero times anything is 0.

When you need to divide a number like 45 by 9, you can rely on basic facts (9 × 5 = 45), but if you need to divide 738 by 9, that's not in the basic facts you've memorized.

The strategy you want to use instead has a logic, a way of thinking about what's going on, and an *algorithm*, a step-by-step process for actually doing it. Algorithms can feel like magic, especially if you don't understand the logic behind them, so let's look at the logic first.

 **DEFINITION**

An **algorithm** is a list of steps necessary to perform a process.

The number 738 is made up of 7 hundreds, 3 tens, and 8 ones. Think of them like paper money. You want to divide by 9. Look first at the hundreds. Can you deal out 7 hundred dollar bills into 9 piles? Not without leaving some piles empty, because there are 9 piles but only 7 hundreds. So exchange all the hundred dollar bills for ten dollar bills. 7 hundreds give you 70 tens, and the 3 tens you already had make 73 tens.

Can you deal the 73 tens out into 9 piles? You could deal 8 tens into each pile and have 1 ten left over. Okay, each pile has 8 tens. Take the extra 1 ten, trade it for 10 ones, and add on the 8 ones you already had. You've got 18 ones. Deal them out into the 9 piles, and each pile will get 2 ones. Each pile got 8 tens and 2 ones. That's 82. 738 ÷ 9 = 82.

No one wants to think about all that dealing out and exchanging every time there's a problem to be done, especially if the numbers are large. That's why there's an algorithm. Let's look at the same problem with the algorithm. We set up the division problem with the divisor outside and the dividend, the number that's being divided up, inside the long division symbol, or division bracket. As you complete the problem, the quotient will be written on top of the bracket.

$$9\overline{)738}$$

Take the process digit by digit. How many 9s are there in 7? None, so you can put a zero over the top of the 7. Multiply the digit you just put in the quotient, 0, times the divisor, 9, and put the result under the 7. Subtract, and check that the result of the subtraction is smaller than the divisor. 7 is less than 9, so move on.

$$9\overline{)738} \\ \underline{\phantom{00}0} \\ \phantom{00}7$$

$$\begin{array}{r} 0\phantom{00} \\ 9\overline{)738} \\ \underline{0}\phantom{00} \\ 7\phantom{00} \end{array}$$

Those 7 hundreds change to 70 tens and get added to the 3 tens to make 73 tens. Show that by bringing the 3 down. Divide 9 into 73. It goes 8 times, so put an 8 up in the quotient. Multiply that 8 times the divisor, put the product (72) under the 73, and subtract. Check that the result is less than the divisor.

$$\begin{array}{r} 08\phantom{0} \\ 9\overline{)738} \\ \underline{0}\phantom{00} \\ 73\phantom{0} \\ \underline{72}\phantom{0} \\ 1\phantom{0} \end{array}$$

That's the one leftover ten, which changes to 10 ones. Bring down the 8 to make 18 ones and divide 9 into 18. Put the 2 in the quotient, multiply 2 times the divisor, put the 18 under the 18 and subtract.

$$\begin{array}{r} 082 \\ 9\overline{)738} \\ \underline{0}\phantom{00} \\ 73\phantom{0} \\ \underline{72}\phantom{0} \\ 18 \\ \underline{18} \\ 0 \end{array}$$

738 ÷ 9 = 82. You can drop that zero in the front, and in the future, you can just leave a blank space if the first digit in the quotient is a zero. Don't drop any zeros that come later, though. They're important place holders.

Long division doesn't always work out to be as tidy as that last example. Sometimes you'll get to the end of the dividend, with nothing left to bring down, but you'll have something left over. For example, if you divide 49 ÷ 6, you'll get a quotient of 8, but there will be 1 left over, because 6 × 8 is only 48. The leftover 1 is called the *remainder*. For now, just say you have a remainder. Later on, you'll see other ways to handle it.

 **DEFINITION**

A **remainder** is the number left over at the end of a division problem. It's the difference between the dividend and the product of the divisor and quotient.

Try a few problems to be sure you've mastered the algorithm.

**CHECK POINT**

Use the long division algorithm to find each quotient.

26. 4,578 ÷ 42

27. 3,496 ÷ 19

28. 16,617 ÷ 29

29. 681 ÷ 14

30. 1,951 ÷ 35

## The Least You Need to Know

- Addition and subtraction are inverse operations, as are multiplication and division.
- Compatible numbers are number pairs that add up to ten. Looking for compatible numbers when adding and subtracting can allow you to complete the operation more quickly.
- When performing operations with multi-digit numbers, keep the place value columns aligned.
- The long division algorithm is a process that allows you to divide large numbers. To use this algorithm, remember: divide, multiply, subtract, compare, bring down, and repeat.

# Order of Operations and Integers

In the previous chapter, we focused on arithmetic that used whole numbers. The whole numbers are the counting numbers (1, 2, 3, 4, 5, and so on) and zero, or the set {0, 1, 2, 3, 4, 5 ...}. It's time now to look at some of the rules about how you should approach more complicated problems. In this chapter, we'll look at the order of operations and at what people do when they don't want you to follow those rules.

You've already encountered the commutative property and the associative property, the rules that let you rearrange a problem that's all addition or all multiplication. Because your problems aren't always one operation, it's time to meet the distributive property, which will give you some options for dealing with addition and multiplication in the same problem and help with mental math along the way.

The set of whole numbers may contain infinitely many numbers, but even the whole numbers aren't big enough to express all the ideas people have about numbers, so in this chapter you'll get acquainted with a set of numbers called the integers. These positive and negative numbers let you express ideas of opposites and give you a way to answer questions you might have once been told were impossible.

## In This Chapter

- The rules for the order of operations
- Using grouping symbols to make tasks clearer
- Applying the distributive property
- Using negative numbers to represent opposites
- Finding absolute value
- Performing operations with integers

# Order of Operations

What does $2 + 3 \times 7$ equal? Some people might say $2 + 3$ is 5 and 5 times 7 is 35. Those people started at the left and did what they saw as they saw it, moving across the line, a logical enough approach. Other people would do the multiplying first and say that's 21, then add on the 2 to get 23. Those people might have thought that multiplication is repeated addition, so do that first, then the simple addition.

And what about the order of the numbers? You learned that addition is associative and commutative, so you can rearrange an addition problem. And you can rearrange a multiplication problem, too, so can you rearrange one that has both operations? Can you change $2 + 3 \times 7$ to $2 + 7 \times 3$? If you do, the multiplication-first people will get the same answer, but the left to right people will get a different one than they got the first time.

Those are great questions, and you certainly don't want $2 + 3 \times 7$ to have two (or more) different values, depending on who does the arithmetic. That wouldn't be practical. So what's the solution?

One way to communicate what should be done first is to add parentheses. The expression $(2 + 3) \times 7$ tells you to add first, then multiply. $(2 + 3) \times 7 = 5 \times 7 = 35$. On the other hand, $2 + (3 \times 7)$ says multiply first, so you get $2 + (3 \times 7) = 2 + 21 = 23$.

That's helpful, but you still need to know what to do, even if people writing the problem don't use parentheses. So there's an agreement among people who do arithmetic about what to do first, second, and so on.

That agreement is called the *order of operations*. What's in parentheses will always get done first, if there are parentheses, and after that, the order will be exponents (think of them like super-multiplication), then multiplication and division, and finally addition and subtraction.

**DEFINITION**

The **order of operations** is an agreement among mathematicians that operations enclosed in parentheses or other grouping symbols should be done first, and then exponents should be evaluated. After that, do multiplication and division as you meet them moving left to right, and finally do addition and subtraction as you meet them, moving left to right.

There are a lot of memory devices to help you remember the order, and one of them, *PEMDAS*, is so common many people use it in place of order of operations, as if it were a name. The letters in PEMDAS are meant to help us remember to:

- P: Simplify expressions inside Parentheses

- E: Evaluate powers, or numbers with Exponents

- MD: Multiply and Divide, moving from left to right

- AS: Add and Subtract, moving from left to right

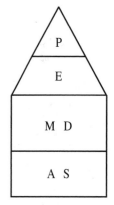

**DEFINITION**

**PEMDAS** is a mnemonic, or memory device, to help you remember that the order of operations is parentheses, exponents, multiplication and division, addition and subtraction.

Some people use a sentence, like Please Excuse My Dear Aunt Sally, to help remember the letters in PEMDAS, and others use an image of a house. The house reminds you that multiplication and division are done together, and so are addition and subtraction.

**MATH TRAP**

Multiplication and division have the same priority. Don't do all the multiplication and then all the division. Do multiplication or division as you meet them as you work across the line. The same is true for addition and subtraction. Do them as you come to them.

Here are some examples of the order of operations at work. Look at each example first, and think about what you would do. Then read the explanation to see if you had all the rules in order.

Example 1: Simplify $(4 \times 5^2 - 8) \times 3$

There are parentheses here, so work on what's inside them first. It's a little problem of its own, so follow the order of operations. First is the exponent. The exponent of 2 on the 5 tells you to multiply $5 \times 5$. That makes the problem $(4 \times 25 - 8) \times 3$. Do the multiplication inside the parentheses, and you have $(100 - 8) \times 3$. Then you can subtract to get $92 \times 3$. That gives you 276.

Ready for a more complicated problem? Try this example.

Example 2: $7 \times 20 - 2 \times 4 + 3^2 + 12 \div 4$

There are no parentheses, so take care of the exponent first. $3^2 = 9$, so the problem becomes $7 \times 20 - 2 \times 4 + 9 + 12 \div 4$. Next, do multiplication and division as you come to them, starting on the left and moving across the line: first $7 \times 20$, then $2 \times 4$, and then $12 \div 4$.

$7 \times 20 - 2 \times 4 + 9 + 12 \div 4$

$= 140 - 2 \times 4 + 9 + 12 \div 4$

$= 140 - 8 + 9 + 12 \div 4$

$= 140 - 8 + 9 + 3$

When all the multiplying and dividing are taken care of, return to the beginning of the line, and do addition and subtraction as you come to them.

$140 - 8 + 9 + 3$

$= 132 + 9 + 3$

$= 141 + 3$

$= 144$

Sometimes you'll see more than one set of grouping symbols in the same problem, and sometimes you may see one set inside another. Different types of grouping symbols are often used to help you tell which is which.

Example 3: $7[120 - 2(4 + 3)^2 + 12] \div 2$

This problem uses parentheses inside of brackets. Brackets have the same meaning as parentheses. They're just a different shape so you don't get confused. Work from the inside out. Notice that there are numbers in front of the parentheses and brackets with no operation sign in between. This is another way of writing multiplication.

Do what's in the parentheses first.

$7[120 - 2(4 + 3)^2 + 12] \div 2$

$= 7[120 - 2(7)^2 + 12] \div 2$

Now work on the problem inside the brackets: $120 - 2(7)^2 + 12$. Follow the order of operations and don't worry about anything else until you finish with this part.

$7[120 - 2(7)^2 + 12] \div 2$

$= 7[120 - 2(49) + 12] \div 2$

$= 7[120 - 98 + 12] \div 2$

$= 7[120 - 98 + 12] \div 2$

$= 7[22 + 12] \div 2$

$= 7[34] \div 2$

Now you can finish up, starting from the left.

$7[34] \div 2$

$= 238 \div 2$

$= 119$

Although a problem may look complicated, using the order of operations can break it down into manageable steps.

**WORLDLY WISDOM**

A number written in front of parentheses or other grouping symbols without an operation sign in between tells you to multiply that number by the result of the work in parentheses. For example, $3(4 - 2) = 3(2) = 3 \times 2 = 6$.

A minus sign in front of parentheses tells you to subtract the result of the parentheses from the number that precedes the minus sign. For example, $10 - (5 + 1) = 10 - (6) = 10 - 6 = 4$.

Try applying the order of operations to some problems.

**CHECK POINT**

Complete each arithmetic problem.

1. $9 - 4 \times 2$

2. $3^2 - 2 \times 4 + 1$

3. $(2^3 - 5) \times 2 + 14 \div 7 - (5 + 1)$

4. $[(2^3 - 5) \times 2 + 14] \div 5 - 3 + 1$

5. $[(3^2 - 2 \times 4) + 1]2 + [11 - 8 + 5(3 + 1)]$

# The Distributive Property

Occasionally, when you're following the order of operations, you many notice that you have a little wiggle room in the rules. For example, at the end of Example 3 above, I said $7[34] \div 2 = 238 \div 2 = 119$, but you might notice that $7(34 \div 2) = 7(17)$ is also 119. That's because dividing the product of 7 and 34 by 2 is equivalent to dividing one of them by 2 and then multiplying. Dividing is so closely related to multiplication that you have a little flexibility. But it's risky to try to cheat the rules. Yes, sometimes in a problem involving only multiplication and division or only addition and subtraction, you might be able to change things around a bit, but most of the time, you'll want to stay with PEMDAS.

There's an important rule in arithmetic that gives you a choice of what to do first when you have multiplication and addition or subtraction, however. That rule is called the *distributive property*. It says that if you're asked to do an addition or subtraction and then multiply the result by a number, you can do just that, or you can choose to spread out the multiplication. Before you add, you multiply each of the addends by the multiplier, like this: $4(5 + 3) = 4 \times 5 + 4 \times 3 = 20 + 12 = 32$. You get the same answer if you do the addition first $4(5 + 3) = 4(8) = 32$. The distributive property gives you the chance to decide which you think will be easier.

> **DEFINITION**
>
> The **distributive property** says that for any three numbers $a$, $b$, and $c$, $c(a + b) = ca + cb$. In other words, the answer you get by first adding $a$ and $b$ and then multiplying the sum by $c$ will be the same as the answer you get by multiplying $a$ by $c$ and $b$ by $c$ and then adding the results.

Consider the problem $17(86 + 14)$.

The easiest way to do this one is to add first.

$17(86 + 14) = 17(100) = 1,700$

Now think about $4(125 + 325)$.

This might be easier if you distribute.

$4(125 + 325) = 4(125) + 4(325) = 500 + 1,300 = 1,800$

> **WORLDLY WISDOM**
>
> The distributive property is a great help in mental math. You might not think you can multiply $45 \times 98$ in your head, but if you think of it as $45(100 - 2)$, it's $4,500 - 90$, which is $4,410$.

When you see a minus sign in front of parentheses, as in $14 - (8 + 3)$, it says to do what's in the parentheses and subtract the result from the number before the minus sign. The distributive property gives you the option to distribute the minus and rewrite the problem as $14 - 8 - 3$. Taking away the sum of 8 and 3 is the same as taking away 8 and taking away 3.

**CHECK POINT**

Tell whether it's easier to perform the operation in parentheses first or to distribute first. Then solve the problem.

6.  $2(35 + 14)$                    9.  $15(40 - 14)$

7.  $3(20 + 8)$                    10.  $250(1,000 - 400)$

8.  $7(100 - 2)$

By now you know how to do all your arithmetic, and you've memorized your addition facts and multiplication tables. You know that the commutative, associative, and distributive properties are handy helpers, and you've got PEMDAS totally under control. What else is there to say about arithmetic? Well, a lot actually. Some of it can wait for a later chapter, but this is a good time to tackle the half-truth you probably learned about subtraction.

As soon as you begin to subtract, you run into a problem. You can subtract $5 - 3$ and get 2, but what happens if you subtract $3 - 5$? The easy answer is, "you can't take a bigger number away from a smaller one." In some sense that's true. If you only have 3 cookies, you can't give me 5 cookies, which is sad, because I love cookies.

But what if you promised me 5 cookies? You owe me 5 cookies. If you give me 3 cookies, you still owe me 2 more cookies. You don't have cookies, but you owe 2 cookies. In a sense, you have less than no cookies, because even if you get more cookies, you have to give me 2. How do we write, in symbolic form, that you owe me 2 cookies? That opposite-of-having idea is written by putting a negative sign in front of the 2. You have -2 cookies. The number -2, or negative 2, is the opposite of 2.

This notion of needing a way to express opposites leads you to a bigger set of numbers. Every number has an opposite, and the whole numbers didn't take that into account. So you need a larger set of numbers that will include all the whole numbers and their opposites. Those numbers are called the integers.

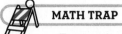

**MATH TRAP**

Try to get in the habit of calling a number less than zero "negative" rather than "minus." The word "minus" signals subtraction, and not everything you do with numbers less than zero is a subtraction. Talking about "negative 4" reminds you that it's the opposite of "positive 4."

# The Integers

We call the set of whole numbers and their opposites *integers*. The set of integers can be written like this: {…, -4, -3, -2, -1, 0, 1, 2, 3, 4, …}. The order in which the numbers are written conveys their value. Any negative number is less than zero. The more you owe, the less you have, so -5 is less than -3. Just as the counting numbers, or positive numbers, go up forever without end, the negative numbers go down forever.

**DEFINITION**

The **integers** are the set of numbers that include all the positive whole numbers and their opposites, the negative whole numbers, and zero.

## The Number Line

One way to visualize the integers is to place them on a *number line*. Usually the line is drawn horizontally, but you could make a vertical number line if you chose. Zero is marked at a point on the line, and then the line is broken into sections of equal length, to the left and the right of zero.

$$\text{-8 -7 -6 -5 -4 -3 -2 -1 } 0 \text{ 1 2 3 4 5 6 7 8}$$

The positive numbers are placed on equally spaced marks to the right of 0, getting larger as you move to the right. The negative numbers are placed on equally spaced marks to the left of 0, with -1 at the first mark left of 0, then -2 at the next mark to the left, and then -3 and so on. The arrows on the ends of the line remind you that the numbers keep going. (We'll fill in the spaces between the integers soon.)

When you compare two numbers, remember that the number to the left is the smaller one. The number 5 is to the left of 9 on the number line, so 5 is less than 9, and you know that -8 is smaller than -1 because -8 is to the left of -1 on the number line. Any negative number is smaller than any positive number, and the negatives are to the left of zero and the positives are to the right.

📖 **DEFINITION**

The **number line** is a line divided into segments of equal length, labeled with the integers. Positive numbers increase to the right of zero, and negative numbers go down to the left.

## Absolute Value

The integers came into being because people wanted a way to express opposite ideas like owing and having, or winning and losing. The number line lets you do that by assigning directions. Positive numbers go to the right (or up, on a vertical line), and negatives go left (or down).

But sometimes you don't really care about the direction. You just want to know "how far?" The *absolute value* of a number is how far from 0 the number is, regardless of direction. The positive number 4 is 4 steps away from 0. The absolute value of 4 is 4. The negative number -4 is also 4 steps away from 0, but in the other direction. Absolute value doesn't care about direction, so the absolute value of -4 is also 4.

📖 **DEFINITION**

The **absolute value** of a number is its distance from zero, without regard to direction. Absolute value cannot be negative.

The symbol for the absolute value of a number is a set of vertical bars that surround the number. The absolute value of -9 is written $|-9|$, so you can write $|-9| = 9$. If you write $|12| = 12$, you're saying that the absolute value of 12 is 12, or that the number 12 is 12 steps away from 0. The absolute value symbols can act like parentheses, so if you see arithmetic inside them, do the arithmetic first, then find the absolute value of the answer.

Find each absolute value.

11. $|-19| =$

12. $|42| =$

13. $|0| =$

14. $|5 - 3| =$

15. $7 + |5 - 3| =$

# Arithmetic with Integers

Arithmetic with integers may be clearer if you have a sense of owing and having, or gaining and losing. You can use the directional sense of the number line, right and left or up and down, to help as well. We'll look at each operation and discuss the rules associated with that operation.

Before we begin, let's address important point about notation. If you write negative numbers with a sign in front, like -7, it would make sense to write positive numbers with a sign in front, like +6. It's perfectly correct to do that, and sometimes you will see it, but most times, you won't. You can assume that a number that doesn't have a sign in front, like 9 or 3, is a positive number, but a negative number, like -5 and -101, will always have a sign to tell you it's negative. A number with a negative sign is also called a signed number.

## Adding Signed Numbers

Adding a positive number to a positive number is nothing new. 4 + 7 = 11, and 73 + 65 = 138. A gain plus a gain is a bigger gain, so a positive number plus a positive number is a positive number. And a loss plus a loss is a loss, so a negative number plus a negative number is a negative number. If you lose $5 and then you lose $8, you've lost a total of $13, so -5 + -8 = -13. To put it in formal language, if you're adding numbers that have the same sign, you add their absolute values and give the answer the same sign as the original numbers.

But what happens when you add numbers with different signs, when you add a positive and a negative? Unfortunately, that's one of the times when the answer is "it depends." If you're playing football and you lose 3 yards on the first play and gain 14 yards on the second, the big gain cancels out the loss and still moves you forward.

Picture it on a number line (or a football field, if you're so inclined). Call the line of scrimmage, your starting point, the point we label 0. The first play is a loss of 3, taking you to -3.

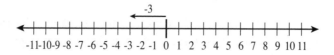

Your second play starts from -3 and moves you 14 spaces in the positive direction. The first 3 of those 14 bring you back to 0, and then you continue for another 11 in the positive direction: -3 + 14 = 11. A negative number plus a positive number gave you a positive number.

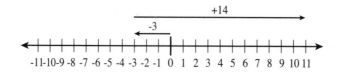

But what if things were reversed and you gained 3 yards and then lost 14 yards? That's 3 + -14. Start from 0 again. A gain of 3 moves you 3 to the right, but then the loss moves you back to the left. You give up the 3 you had gained and keep moving left another 11 spaces. 3 + -14 = -11.

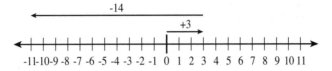

A big gain overpowered a small loss, resulting in a gain, but a big loss overpowered a small gain, leaving you with a loss. In mathematical terms, the number with the larger absolute value dominates the addition and gives its sign to the result. The absolute value of the answer is the difference between the absolute values of those two competing forces.

To add numbers with different signs:

• Subtract the absolute values.

• Take the sign from the number with the larger absolute value.

**CHECK POINT**

Complete each addition problem. Use a number line to help.

16.  -15 + 25

17.  19+ -12

18.  -23 + 14

19.  -58 + -22

20.  147 + -200

## Subtracting Signed Numbers

There's a simple rule for subtracting signed numbers: don't. This doesn't mean you can just ignore those problems. Subtraction, as you saw in the last chapter, is the opposite, or undoing, of addition, and learning a lot of new and separate rules for subtraction is effort you don't need to expend.

When you're asked to subtract, add the opposite. Instead of 12 − 8, which you know equals 4, think of 12 + -8, which also equals 4. Then when you're asked to do -14 − 7, you can just think of -14 + -7 and quickly arrive at -21. 6 − (-3) will become 6 + 3, which is clearly 9. This rule is sometimes referred to as "keep, change, change" because you keep the first number as it is, change to addition, and change the second number to its opposite.

To subtract signed numbers:

- Leave the first number as is.

- Change to addition.

- Change the second number to its opposite.

- Add, following the rules for addition.

**CHECK POINT**

Complete each subtraction by adding the opposite.

21. -17 − 4

22. 39 − 24

23. 26 − -12

24. -83 − 37

25. -48 − -32

## Multiplying and Dividing Signed Numbers

Because multiplication is a shortcut for repeated addition, it's fair to ask if you can use some of the addition rules to help with multiplication of signed numbers, and because division is the opposite or undoing of multiplication, you'd certainly expect to be able to apply some of the same rules about signs. Let's look at multiplication first, and then you'll see that there's very little more to say about division.

Multiplication is really repeated addition. $5 \times 12$ means "add 5 twelves together" (or 12 fives, if you prefer). $5 \times 12 = 12 + 12 + 12 + 12 + 12$, and adding positive numbers gives you a positive number. $5 \times 12 = 60$. Nothing new there.

What if you multiply a positive number by a negative number? What is $3 \times -4$? It means "add 3 copies of the number -4" or "take 3 losses of 4 each." That's $-4 + -4 + -4$, and adding negative numbers gives you a negative. $3 \times -4 = -12$. If you're wondering what to do with a negative times a positive, like $-5 \times 7$, remember that multiplication is commutative. $-5 \times 7 = 7 \times -5 = -35$, and no new rules are necessary.

When you multiply a negative number by a negative number, remember that the negative sign means "the opposite of." This means $-5 \times 7$ is the opposite of $5 \times 7$, or the opposite of 35, or -35. If both numbers are negative, as in $-2 \times -4$, you're asking for the opposite of $2 \times -4$. You know that $2 \times -4$ is -8, and the opposite of that is 8. $-2 \times -4 = 8$. A negative multiplied by a negative is a positive.

To multiply signed numbers, multiply the absolute vales and follow these rules for signs.

Positive × Positive = Positive

Positive × Negative = Negative × Positive = Negative

Negative × Negative = Positive

If you're multiplying more than two numbers, you can save some time by counting the number of negative signs. If the number of negatives is even, the product will be positive. If the number of negatives is odd, the product will be negative. For example, -2 × -5 × -3 × -1 = (-2 × -5) × (-3 × -1) = 10 × 3 = 30. Four negatives, an even number, make a positive. But put one more negative into the problem and you have -2 × -5 × -3 × -1 × -4 = (-2 × -5) × (-3 × -1) × -4 = 10 × 3 × -4= 30 × -4 = -120. With an odd number of negatives, your answer is negative.

> **WORLDLY WISDOM**
>
> A quick way to remember the rules for multiplying integers: multiply same signs, your answer is positive; multiply different signs, it's negative.

> **CHECK POINT**
>
> Complete each multiplication problem.
>
> 26.  -4 × 30                    29.  -11 × -43
>
> 27.  8 × -12                    30.  -250 × 401
>
> 28.  -7 × 15

You'll be pleased to know that the rules for division of signed numbers are the same as the rules for multiplication, except that, obviously, you divide the absolute values instead of multiplying.

To divide signed numbers, divide the absolute values and follow these rules for signs.

Positive ÷ Positive = Positive

Positive ÷ Negative = Negative ÷ Positive = Negative

Negative ÷ Negative = Positive

Following those rules, you can see that 42 ÷ 6 = 7, and 84 ÷ -4 = -21. -15 ÷ 5 = -3, but -15 ÷ -5 = 3.

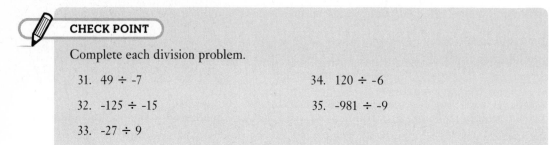

**CHECK POINT**

Complete each division problem.

31.  49 ÷ -7

32.  -125 ÷ -15

33.  -27 ÷ 9

34.  120 ÷ -6

35.  -981 ÷ -9

## The Least You Need to Know

- When completing an arithmetic problem, follow the order of operations: parentheses first, then exponents, then multiplication and division, and lastly, addition and subtraction.

- The distributive property says that when you need to add two numbers and then multiply by a third, you can choose to multiply each of the first two by the third and then add. You'll get the same answer.

- The absolute value of a number is its distance from zero on the number line.

- To add integers with the same sign, add the absolute values and keep the sign.

- To add integers with different signs, subtract the absolute values and take the sign on the number with the larger absolute value.

- To multiply or divide integers, multiply and divide without the signs. Then it the signs of the two numbers are the same, your answer is positive. If the signs are different, your answer is negative.

# Factors and Multiples

In any multiplication problem, the numbers being multiplied are factors and the result of the multiplication is the product. We've used that language in earlier chapters. You can also say that the product is a multiple of each of the factors. It's time to take a step beyond the vocabulary now and look at factors and multiples in more detail. It may seem odd to give a whole chapter to multiplication, but ideas that have to do with factors and multiples come up in so many situations that it's wise to take the time to look at all of them.

In this chapter, we'll look at numbers that have many factors and those that have few. You'll practice different ways to find all the possible factors of a number and work at expressing a number as the product of factors, using the smallest possible factors. Exponents will show up again to help you write those factors efficiently. All this talk about factors will lead to two important ideas: the greatest common factor and the least common multiple of two or more numbers.

## In This Chapter

- How to recognize multiples of a number
- Finding factors of a number
- Identifying prime and composite numbers
- Prime factorization and how to find it

# Prime Numbers

When you multiply 1 by any number, the product is that number. 1 × 4 = 4, 1 × 17 = 17, and 1 × 132 = 132. The number 1 is a factor of any number. When you write a multiplication problem like those, with 1 as a factor, the other factor is always the same as the product. So if 1 is a factor of 4, then 4 is also a factor of 4, and that's the case for any number. Any number × 1 = that number. The number 1 is a factor of every number, and every number is a factor of itself. Every number can be written as itself times 1.

All that multiplying by 1 may seem boring, and many numbers have other factors that make more interesting problems. The number 6, in addition to being 1 × 6 is also equal to 2 × 3. And some numbers have lots of possible factors. 24 could be 1 × 24, 2 × 12, 3 × 8, or 4 × 6. But there are numbers that only have one set of factors. Numbers whose only possible factors are themselves and 1 are called *prime numbers*. Numbers that have other possible factors besides themselves and 1 are called *composite numbers*. You might think that only small numbers like 2 and 3 are prime numbers, but that's not the case.

> **DEFINITION**
>
> A **prime number** is a whole number whose only factors are itself and 1. A **composite number** is a whole number that is not prime because it has factors other than itself and 1.

You can use blocks to help you visualize this concept. If you have a prime number of blocks, there's only one rectangle you can make. Try to rearrange and you just can't get a rectangle.

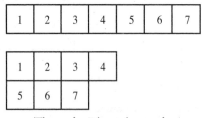

*The number 7 is a prime number.*

With a composite number of blocks, you can always make at least two rectangles.

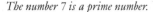

*The number 8 is a composite number.*

**MATH TRAP**

Most people think 1 is a prime number. Most people are wrong. It's understandable. The number 1 has no other factors, but it doesn't quite fit the definition of prime. Its factors aren't itself and 1. They're itself and itself, or 1 and 1. It's sort of strange, but it gets its own category. The number 1 is neither prime nor composite. It's called a unit.

## Finding Prime Numbers

Take a look at the following chart, which shows the whole numbers 1 through 50. You know 2 is prime, because its only factors are 2 and 1, but all the other even numbers have 2 as a factor, so they're composite. Multiples of 2 can be blocked out.

The number 3 is prime, but all multiples of 3 are composite, and 5 is prime, but any other number that ends in 5 is composite because it's divisible by 5. All the multiples of 3 or 5 that haven't already been eliminated can be blocked out.

The numbers 4 and 6 have already been marked as composite, but 7 is prime. Several multiples of 7 have already been eliminated as composite. You just need to mark 7 × 7, or 49, as composite, and believe it or not, everything left is prime, except for 1, which is in a class by itself. That means 16 of the first 50 counting numbers are prime.

| 1 | 2 | 3 | 4 | 5 | 6 | 7 | 8 | 9 | 10 |
|----|----|----|----|----|----|----|----|----|----|
| 11 | 12 | 13 | 14 | 15 | 16 | 17 | 18 | 19 | 20 |
| 21 | 22 | 23 | 24 | 25 | 26 | 27 | 28 | 29 | 30 |
| 31 | 32 | 33 | 34 | 35 | 36 | 37 | 38 | 39 | 40 |
| 41 | 42 | 43 | 44 | 45 | 46 | 47 | 48 | 49 | 50 |

**WORLDLY WISDOM**

As the numbers get bigger, prime numbers become fewer and more spread out, but mathematicians are still finding them. The largest known prime number is $2^{57,885,161} - 1$, a number with 17,425,170 digits.

How do you know if a large number is prime? You won't need to worry about extremely large numbers with thousands of digits, but you may need to know about numbers in the hundreds or thousands. Is 136 prime? How about 2,485? What about 901?

Sometimes it's easy to tell. The number 136 can't be prime because it's even, so it has 2 as a factor as well as itself and 1. In the same way, 2,485 ends in 5, so it's divisible by 5 and therefore not prime. The number 901 is harder. It's not even, and it's not a multiple of 5, because it doesn't end in 5 or 0, but what about other numbers? Does it have any divisors?

There are a few shortcuts, like recognizing even numbers (numbers that end in 0, 2, 4, 6, or 8) and multiples of 5 (numbers that end in 5 or 0). You can check if a number is divisible by 3 by adding the digits. If the sum is divisible by 3, so is the original number. If you get a large number as the sum of the digits and you don't know if it's divisible by 3, you can add the digits again. You can add the digits as many times as necessary until you can decide if the sum is a multiple of 3.

For example, to see if 45,829 is a multiple of 3, add 4 + 5 + 8 + 2 + 9 = 28. Is 28 a multiple of 3? It's not, but if you're not sure, add 2 + 8 = 10. You know 10 is not a multiple of 3, so 45,829 is not a multiple of 3.

The same idea will work for 9. If you add the digits and get a multiple of 9, the original number is divisible by 9. Here's a chart of the tests you might want to remember.

## Tests for Divisibility

| Multiple of: | Test | Example |
|---|---|---|
| 2 | Ends in 2, 4, 6, 8, or 0. | 45,938 is a multiple of 2 because it ends in 8. 673 is not a multiple of 2. |
| 3 | Sum of the digits is divisible by 3. | 9,348 is a multiple of 3 because 9 + 3 + 4 + 8 = 24, which is a multiple of 3. (Check 2 + 4 = 6, a multiple of 3.) |
| 4 | The last two digits are a multiple of 4. | 35,712 is a multiple of 4 because the last two digits, 12, is a multiple of 4. |
| 5 | Ends in 5 or 0. | 74,935 and 680 are multiples of 5. |
| 6 | If it's a multiple of 2 and a multiple of 3, it's a multiple of 6. | 65,814 is a multiple of 2 because it ends in 4. It's a multiple of 3 because 6 + 5 + 8 + 1 + 4 = 24, a multiple of 3. So 65,814 is a multiple of 6. |
| 8 | The last 3 digits are a multiple of 8. | 42,864 is a multiple of 8 because 864 is a multiple of 8. |
| 9 | Sum of the digits is a multiple of 9. | 39,348 is a multiple of 9, because 3 + 9 + 3 + 4 + 8 = 27, a multiple of 9. |
| 10 | Ends in 0. | 483,940 is a multiple of 10. |

Sometimes, to find out if a number has any factors, or divisors, other than itself and 1, you have no choice but to just divide by each of the primes. The number 901 is not even, so 2 is not a factor. Its digits add to 10, so it's not divisible by 3. It doesn't end in 5 or 0, so it's not a multiple of 5. But you just have to divide by 7 and see what happens. 901 ÷ 7 = 128 with 5 left over, so 7 is not a factor. 11 is the next prime, but 901 ÷ 11 is 81 with 10 left over, so 11 is not a factor.

But how will you know when you've checked enough prime numbers? One way to tell is by comparing the quotient to the divisor. If the quotient is equal to or smaller than the divisor, you can stop. You've checked enough. And if you get to the point where the quotient is less than or equal to the divisor and you haven't found a factor, your number is prime.

901 ÷ 11 is 81 with 10 left over, and the quotient of 81 is larger than the divisor of 11, so this one isn't done yet.

901 ÷ 13 = 69 with 4 left over, and the quotient of 69 is larger than the divisor of 13, so we keep going.

901 ÷ 17 = 53 exactly, which means 901 is a composite number. It can be written as 17 × 53 as well as 901 × 1.

## Consider Square Numbers

When you multiply any integer by itself, the result is a *square number*. The expression comes from the fact that a square has the same length as width, so when you find its area, you multiply a number by itself. The number 16 is a square number because it's equal to $4^2$ or 4 × 4, and that's the area of a square with a side 4 units long.

 **DEFINITION**

> **Square numbers** are numbers created by multiplying a number by itself, or raising it to the second power. For example, 36 is a square number. It can be written as $6^2$ or 6 × 6.

If you know the square numbers, it will help you see how far you have to go when checking whether a number is prime. 30 × 30, or $30^2$, is 900, which is very close to 901. The biggest prime you'll have to try is the one just above that, or 31. If you know your square numbers, it's easy to know how many primes you need to divide by before declaring a number prime. If you want to know if 147 is prime, think about square numbers near 147. The number 144 is $12^2$, so if you go to the next prime after 12, that's enough. Once you divide 147 by 13, if you haven't found a factor, you can say 147 is prime.

Suppose you want to know if 67 is prime. If you know that $8^2$ is 64, you only have to check 2, 3, 5, 7, and maybe 11. The number 67 is not even, so it's not divisible by 2. Its digits add to 13, so it's not divisible by 3. It doesn't end in 5 or 0, so it's not a multiple of 5. When you divide 67 by 7, you get 9 with 4 left over, so 7 is not a factor. 67 ÷ 11 is 6 (smaller than the divisor) but with 1 left over. 67 is a prime.

**CHECK POINT**

Decide if each number is prime or composite. If it's composite, give a pair of factors other than the number and 1.

1. 51                    4. 229
2. 91                    5. 5,229
3. 173

# Prime Factorization

Factoring is the process of starting with the product and finding the factors that produced it. (Factoring is related to division, but not quite the same.) You could factor 24 by writing it as $2 \times 12$, or $3 \times 8$, or $2 \times 3 \times 4$, and other possibilities.

A number can have lots of different factorizations, but it has only one *prime factorization*. A prime factorization expresses the number as a product of only prime numbers. The prime factorization of 24 is $2 \times 2 \times 2 \times 3$.

**DEFINITION**

The **prime factorization** of a number is a multiplication that uses only prime numbers and produces the original number as its product.

You can find the prime factorization just by finding primes that are factors of the number by trial and error, but having a way to organize the search does make things faster. One method for finding the prime factorization of a number is called a factor tree.

## Factor Trees

A *factor tree* is a device that allows you to start with any two factors of a number and to work in an organized fashion to the prime factorization.

**DEFINITION**

A **factor tree** is a method of finding the prime factorization of a number by starting with a pair of factors and then factoring each of those numbers, continuing until no possible factoring remains.

Let's look at the prime factorization of 180 as an example. Start with any pair of factors of 180 that you can recall, maybe 18 and 10. Put 180 at the top of the tree and make a branch for 18 and a branch for 10.

Focus on the 18 branch. Can you find factors for 18? 2 and 9 will work. Add them as branches under 18.

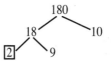

Is 2 a prime? Yes, so mark it with a circle or a box. That branch is done. Is 9 prime? No, but it factors to 3 × 3, and 3 is prime. The tree needs branches under the nine.

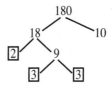

When that branch is down to primes, turn to the 10 branch. 10 = 2 × 5, and both of those numbers are prime.

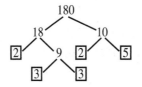

Collect all the primes you marked and you have the prime factorization of 180.
180 = 2 × 2 × 3 × 3 × 5.

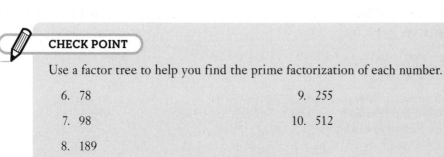

**CHECK POINT**

Use a factor tree to help you find the prime factorization of each number.

6. 78

7. 98

8. 189

9. 255

10. 512

## Using Exponents

Some prime factorizations are easy to write. The prime factorization of 42 is $2 \times 3 \times 7$, which is short and clear. However, the prime factorization of 128 is another story. $128 = 2 \times 2 \times 2 \times 2 \times 2 \times 2 \times 2$. It uses easy numbers, but it uses so many of them that it can get confusing. Is that six 2s or seven? It's easy to make a mistake.

That's why you may want to use exponents to condense the prime factorization and make it easier to understand. The prime factorization of $40 = 2 \times 2 \times 2 \times 5$, but it could be written as $40 = 2^3 \times 5$. The exponent of 3 says to use 2 as a factor 3 times. With that notation, you can write the prime factorization of 128 as $128 = 2^7$ and know for certain that it was seven 2s.

**CHECK POINT**

Use exponents to write the prime factorization of each number in a compact form.

11.  200

12.  168

13.  672

14.  2,205

15.  22,000

Finding the prime factorization of a number can be an interesting undertaking. It can be difficult, but there's a satisfaction in getting a large number broken down to a product of primes. The truth, however, is that finding a prime factorization is not usually your final goal, but a tool to help you find two other very useful numbers: the *greatest common factor* and the *least common multiple*.

# Greatest Common Factor

The greatest common factor, or GCF, of two numbers is the largest number that is a factor of both. The greatest common factor of 21 and 24 is 3. The factors of 21, other than 1 and 21, are 3 and 7. The factors of 24 are 1, 2, 3, 4, 6, 8, 12, and 24. The largest number on both lists is 3.

**DEFINITION**

The **greatest common factor (GCF)** of two numbers is the largest number that is a factor of both.

To find the greatest common factor of 70 and 105, first find the factors of 70. The factors of 70 are 1, 2, 5, 7, 10, 14, 35 and 70. Then find the factors of 105. Those are 1, 3, 5, 7, 15, 21, 35, and 105. The largest number that shows up on both factor lists is 35, so 35 is the GCF of 70 and 105.

There are times when one of the numbers is the GCF of the two numbers. The GCF of 8 and 32 is 8, which is a factor of both 32 and of 8, because every number is a factor of itself. The factors of 8 are 1, 2, 4, and 8, and the factors of 32 are 1, 2, 4, 8, 16, and 32. 8 is the largest number that appears on both lists.

At the other extreme, sometimes the only factor the two numbers will have in common is 1. Remember 1 is a factor of every number. Numbers that have only 1 as a common factor are called *relatively prime.* The numbers 12 and 49 are relatively prime because the factors of 12 are 1, 2, 3, 4, 6, and 12, and the factors of 49 are 1, 7, and 49. The only number on both lists is 1.

**DEFINITION**

Two numbers are **relatively prime** if the only factor they have in common is 1.

For small numbers, making a list of factors isn't difficult. If the number is small, the list of factors can't be too long. But larger numbers can have long factor lists, so it's important to have a more efficient method of finding the greatest common factor. That's when you want to turn to the prime factorization.

To find the greatest common factor of 128 and 144, first work out the prime factorization of each number. Factor trees may help you factor to primes. The prime factorization of 128 is seven twos.

$128 = 2^7 = \underline{2 \times 2 \times 2 \times 2} \times 2 \times 2 \times 2$

The prime factorization of 144 has both twos and threes.

$144 = 2^4 \times 3^2 = \underline{2 \times 2 \times 2 \times 2} \times 3 \times 3$

Once you have both prime factorizations, look for any primes that appear in both (these numbers are underlined). It's okay to have repeats if the factor repeats in both factorizations. In this case, there are four 2s that appear in both factorizations. The GCF of 128 and 144 is $2 \times 2 \times 2 \times 2$, which equals $2^4$ or 16.

You can find the GCF of 210 and 495 by starting with the prime factorization of each number.

$210 = 2 \times 3 \times 5 \times 7$

$495 = 3 \times 3 \times 5 \times 11$

Both factorizations have a 3 and a 5. There's only one 3 that's common to both, so only one 3 goes into the prime factorization. 495 has a second 3, but 210 doesn't share it. The GCF of 210 and 495 is $3 \times 5$, or 15.

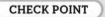

**CHECK POINT**

Find the greatest common factor of each pair of numbers.

16. 18 and 42                    19. 630 and 945

17. 42 and 70                    20. 286 and 715

18. 144 and 242

# Least Common Multiple

If you attend a baseball game at which every twelfth person admitted to the stadium receives a free key ring and every twentieth person gets a free t-shirt, where should you be in line if you want both? You want a spot that's a multiple of 12, so that you get the key ring, but you want it to also be a multiple of 20, so you snag a t-shirt. You also don't want to be too far back in line, so the smallest number that does both jobs is what you're after. Many people would look to be #240 in line because that's 12 × 20, but you can move, if not to the head of the line, at least farther up.

The *least common multiple*, or LCM, of two (or more) numbers is the smallest number that is a multiple of both. If the LCM is a multiple of both numbers, then both numbers divide the LCM, and both are factors of the LCM.

**DEFINITION**

The **least common multiple** of two or more numbers is the smallest number that has each of the numbers as a factor.

The least common multiple of 12 and 20 is much smaller than 240. 240 is a multiple of 12 and of 20. It is a common multiple, but it's not the least, or smallest, common multiple. If you look at lists of the multiples of 12 and 20, you'll see that 60 is the smallest number to show up in both lists.

Multiples of 12: 12, 24, 36, 48, <u>60</u>, 72 …

Multiples of 20: 20, 40, <u>60</u>, 80 …

You can stake out the 60th spot in line at the game, get a good seat, and go home with the key ring and the t-shirt.

Making a list of the multiples of each number and looking for the first number to pop up on both lists is actually a workable method of finding the least common multiple, if the numbers aren't too big. For large numbers, however, it could take a very long time and a lot of multiplying.

Instead, you'll want to start by finding the prime factorization of both numbers. Let's start with an example with small numbers and then try another with larger numbers.

Suppose you need to find the least common multiple of 12 and 15. Find the prime factorization of each number.

12 = 2 × 2 × 3

15 = 3 × 5

Identify any common factors. In this example, 12 and 15 both have a factor of 3. Build the LCM by starting with the common factor or factors, then multiplying by the other factors of 12 and 15 that are not common. 12 and 15 have 3 in common, so that goes into the LCM just once, and then you collect the other factors of 12, which are 2 × 2, and the other factor of 15, the 5.

LCM = 3 × 2 × 2 × 5

The 3 is the common factor, the two 2s come from the factorization of 12, and the 5 is from the 15. Multiply to find that the LCM is 60. 60 is 12 × 5 and 15 × 4, and it's the smallest multiple 12 and 15 share.

Ready to find the least common multiple of 98 and 168? Start by finding the prime factorization of each number. You might want to use a factor tree.

98 = 2 × 7 × 7

168 = 2 × 2 × 2 × 3 × 7

The two numbers have a 2 and a 7 in common, so start building the LCM with those.

LCM = 2 × 7 × ?

You need another 7 from the 98 and another two 2s and a 3 from the 168.

LCM = 2 × 7 × 7 × 2 × 2 × 3 = 1,176

The multiplication for that one might be challenging, but building lists of multiples up into the thousands wouldn't have been an efficient method.

**WORLDLY WISDOM**

If two numbers are relatively prime, their product will be their LCM.

**CHECK POINT**

Find the least common multiple of each pair of numbers.

21.  14 and 35                  24.  21 and 20

22.  45 and 105                 25.  88 and 66

23.  286 and 715

## The Least You Need to Know

- A prime number has no factors other than itself and one.
- A number that is not prime is composite, except 1, which is in a category all its own.
- The prime factorization of a number writes the number as a product of primes.
- The greatest common factor of two numbers is the largest number that is a factor of both.
- The least common multiple of two numbers is the smallest number that is a multiple of both.

# Fractions

Your first encounter with the world of numbers was the simple act of counting. That world is larger than you could have imagined at that first encounter, and over the last few chapters, we've moved from the counting numbers, to the whole numbers, to the integers. When we worked with the integers, the positive and negative whole numbers, I suggested a number line as a way to help you think about them, and I promised to fill in the spaces between the integers soon. That time has come.

The spaces between the integers are filled with numbers that talk about parts of a whole. In this chapter, you'll look at one of the two ways of representing those numbers that show parts. You'll examine the different methods of arithmetic for fractions and learn to add, subtract, multiply, and divide fractions and mixed numbers.

## In This Chapter

- How to find equivalent fractions
- How to find common denominators
- Adding and subtracting fractions and mixed numbers
- Multiplying and dividing fractions and mixed numbers

# The Rational Numbers

The counting numbers, or natural numbers, are the numbers you use to count. When you find that you need a symbol for nothing and add 0 to the counting numbers, you form the set called the whole numbers. When you introduce negative numbers, you have the integers. Now you're turning your focus to numbers that represent parts of a whole, so what language do you use to describe all of this?

*Fractions* are the names our number system gives to symbols representing parts of a whole. A fraction is what we produce when we break a whole number into parts. *Common fraction* is the name we give to a way of writing a fraction as one number divided by another. The number $\frac{1}{2}$ is an example of a common fraction, although usually it's just called a fraction.

 **DEFINITION**

A **fraction** is a symbol that represents part of a whole. Decimal fractions are written in the base ten system, with digits to the right of the decimal point. **Common fractions** are written as a quotient of two integers.

The set of numbers that includes all the integers plus all the fractions is called the *rational numbers*. The name rational comes from ratio, another name for a fraction or a comparison of two numbers by dividing. The rational numbers include any number you can write as a quotient of two integers. So $\frac{1}{2}$ is a rational number, but so is 2, because you can write it as $\frac{2}{1}$. Other examples of rational numbers are $\frac{-4}{13}$, $\frac{163}{71}$, and 0.

**DEFINITION**

**Rational numbers** are the set of all numbers that can be written as the quotient of two integers.

## Proper Fractions and Improper Fractions

When we talk about fractions, we are usually talking about rational numbers written as a quotient of two integers. When a fraction is written this way, the top number is called the *numerator*, and the bottom number is called the *denominator.*

**DEFINITION**

The **denominator** of a fraction is the number below the division bar. It tells how many parts the whole was broken into, or what kind of fraction you have. The **numerator** is the number above the bar, which tells you how many of that sort of part you actually have. In the fraction $\frac{1}{2}$, 2 is the denominator and 1 is the numerator.

Because a whole number like 7 is a rational number, even though you wouldn't usually think of it as a fraction, it can be written as the quotient of two integers. You could write $\frac{7}{1}$, which you would think of as a fraction. You could actually write many different quotients that equal 7, like $\frac{42}{6}$ or $\frac{700}{100}$. Those look like the other quotients you think of as fractions, like $\frac{5}{8}$, $\frac{17}{3}$, and $\frac{4}{51}$.

In those examples, there were two *proper fractions*, or fractions whose value is less than 1. Those were $\frac{5}{8}$ and $\frac{4}{51}$. You can spot a proper fraction because the numerator is less than the denominator. On the other hand, $\frac{17}{3}$ and $\frac{7}{1}$ (and all the fractions equal to $\frac{7}{1}$) are *improper fractions*. They're worth more than 1, because their numerators are larger than their denominators.

> **DEFINITION**
>
> A **proper fraction** is one whose value is less than one. An **improper fraction** is one whose value is more than one.

Fractions use their denominator to tell you how many parts the whole was broken into and their numerator to tell you how many of those parts you have. $\frac{5}{8}$ says the whole was broken into 8 parts (called eighths), and you have 5 of them. That's less than the whole, so it's a proper fraction.

To interpret $\frac{17}{3}$, you have to imagine that many wholes were each divided into 3 parts, called thirds, and you have a total of 17 of those parts. You could piece your parts back together and make 5 full wholes, with 2 of the pieces left over. You can write that as $\frac{17}{3} = 5\frac{2}{3}$, which is called a *mixed number*. Mixed numbers are a combination of a whole number and a fraction, a condensed way of saying $5 + \frac{2}{3}$.

> **DEFINITION**
>
> A **mixed number** is made up of a whole number and a fraction, written side by side. It tells you that you have a number of wholes plus part of another.

**CHECK POINT**

1. Change $\frac{18}{5}$ to a mixed number.

2. Change $\frac{37}{4}$ to a mixed number.

3. Change $7\frac{1}{3}$ to an improper fraction.

4. Change $12\frac{3}{4}$ to an improper fraction.

5. Change $11\frac{7}{8}$ to an improper fraction.

# Arithmetic with Fractions

When you work with fractions, the methods of whole number arithmetic are no longer enough to handle the job. You'll need some new strategies for multiplying and adding. Once you have learned those methods, however, subtraction and division will just be variations and so not much more to learn. In fact, we'll cover addition and subtraction together and multiplication and division together, because they have so much in common. Before that, however, you need to master the key to working happily with fractions: the art of disguise.

## Equivalent Forms

The key to working successfully with fractions is to be able to find equivalent forms of the same fraction. All that means is that you can change the way a fraction looks without changing its worth. The simplest way to do that is to multiply by 1, in disguise.

Disguise? Well, the number 1 can be written as a fraction in many different ways: $\frac{4}{4}$, $\frac{19}{19}$, $\frac{73}{73}$, or any number over itself. Multiplying a number by one will never change its value, but multiplying by 1 wearing the right disguise will change the number's appearance.

You already know how to multiply by 1 when it appears as a whole number. In fact, you know that there is nothing to do when multiplying by 1, because multiplying a number by 1 doesn't change it. But when the 1 puts on its fraction disguise, things are a little different. If, for example, you multiply $\frac{1}{2}$ by 1, you know you should get a number equal to $\frac{1}{2}$. But if the 1 is disguised as $\frac{5}{5}$, the answer won't look like $\frac{1}{2}$, even though that's what it will be worth.

There are lots of numbers that are worth $\frac{1}{2}$. That fraction says the whole was divided into two parts, and you have one of them. If 2 wholes were each divided into two parts, there would be a total of 4 pieces. When the first whole is broken in 2, you should get 1 piece, and then when the second whole is broken, you should get 1 piece of that. You get 2 of the 4 pieces. $\frac{1}{2}=\frac{2}{4}$. There are lots of other fractions that are equal to $\frac{1}{2}$, and in fact, there are lots of names for any fraction. They all have the same value but different appearances.

You don't want to break things into parts and count every time you need a fraction to change its appearance, so there's a simple shortcut. For that shortcut, you need a little preview of fraction arithmetic. The basic rule for multiplying fractions says to multiply numerator times numerator and denominator times denominator. When you want to change the appearance of a fraction, multiply it by a disguised 1. $\frac{1}{2}\times\frac{5}{5}$ is still worth $\frac{1}{2}$ because you're multiplying by 1, but its appearance changes to $\frac{5}{10}$. In the same way, $\frac{1}{2}\times\frac{3}{3}=\frac{3}{6}$ and $\frac{1}{2}\times\frac{19}{19}=\frac{19}{38}$. The same fraction can have many different looks.

You can go the other way, too, taking a fraction to a simpler appearance. This is officially called simplifying, but you'll often hear it called reducing to lowest terms, or just reducing.

**MATH TRAP**

The expression "reducing a fraction" is a dangerous way to describe simplifying. It makes it sound like the fraction is getting smaller, but it's not. The appearance of the fraction is changing, but its value is not. If you must use "reduce"—and most people do—remember that the fraction hasn't changed. $\frac{3}{5}$ may look smaller than $\frac{243}{405}$, but it's the same number.

The fraction $\frac{16}{36}$ can be written in a much simpler form. Find a number that divides both the numerator and denominator, preferably the GCF. In this case, that's 4. Think of the fraction with the numerator and denominator factored: $\frac{16}{36} = \frac{4 \times 4}{9 \times 4}$. Can you see the disguised 1 in there? $\frac{16}{36} = \frac{4 \times 4}{9 \times 4} = \frac{4}{9} \times \frac{4}{4} = \frac{4}{9} \times 1 = \frac{4}{9}$. The fraction $\frac{4}{9}$ is a simpler name for $\frac{16}{36}$.

If you don't know, or don't want to look for, the GCF of the numerator and denominator, you can start the process with any number that divides both. Divide the numerator and denominator by the same number. Repeat until there are no other numbers that will divide both. It may take a while, but you'll get there. If you start with the GCF, you'll only need one try.

**CHECK POINT**

6.  Change $\frac{1}{5}$ to a fraction with a denominator of 40.

7.  Change $\frac{3}{4}$ to a fraction with a denominator of 28.

8.  Change $\frac{5}{7}$ to a fraction with a denominator of 147.

9.  Simplify $\frac{35}{40}$.

10. Simplify $\frac{63}{84}$.

## Multiplying and Dividing Fractions

Since you've already had a preview of the rule for multiplying fractions, let's start the conversation about arithmetic there. The basic rule for multiplication of fractions calls for multiplying numerator times numerator and denominator times denominator, then simplifying if possible. To multiply $\frac{1}{9} \times \frac{3}{7}$, multiply $1 \times 3$ and $9 \times 7$. This gives you $\frac{3}{63}$, which simplifies to $\frac{1}{21}$, because $\frac{3}{63} = \frac{1 \times 3}{21 \times 3} = \frac{1}{21} \times \frac{3}{3} = \frac{1}{21} \times 1$.

Dividing fractions introduces only one extra piece to the multiplication rule. To divide fractions, invert the divisor and multiply. The divisor is always the second fraction. Invert only the divisor, never the first fraction. This inverted, or flipped, version of the fraction is called its *reciprocal*. To divide $\frac{1}{8} \div \frac{3}{4}$, realize that the fraction $\frac{3}{4}$ is the divisor. Invert $\frac{3}{4}$ to get $\frac{4}{3}$, then multiply and simplify: $\frac{1}{8} \times \frac{4}{3} = \frac{4}{24} = \frac{1}{6}$.

> **DEFINITION**
>
> Two numbers are **reciprocals** if when they are multiplied their product is 1. The reciprocal of the whole number 4 is the fraction $\frac{1}{4}$. The reciprocal of a fraction is another fraction with the numerator and denominator switched. $\frac{3}{8}$ and $\frac{8}{3}$ are reciprocals. Each number is the reciprocal of the other.

Much of the work of multiplying and dividing fractions can be made easier by canceling before multiplying. *Canceling* is dividing a numerator and a denominator by the same common factor. You can think of it as simplifying before you multiply instead of after.

> **DEFINITION**
>
> **Canceling** is the process of simplifying a multiplication of fractions by dividing a numerator and a denominator by a common factor.

When you multiply $\frac{8}{9} \times \frac{7}{12}$ you multiply the numerators and multiply the denominators, so your product, before you actually get the multiplying done, is $\frac{8 \times 7}{9 \times 12}$. After you do the multiplying, it's $\frac{56}{108}$, and that needs to be simplified. To simplify it, you'll factor: $\frac{56}{108} = \frac{14 \times 4}{27 \times 4}$. But look back to $\frac{8 \times 7}{9 \times 12}$ for a moment. Can you see that you could factor this version? And because the numbers are smaller, it might be easier. $\frac{8 \times 7}{9 \times 12} = \frac{4 \times 2 \times 7}{9 \times 3 \times 4} = \frac{2 \times 7 \times 4}{9 \times 3 \times 4} = \frac{2 \times 7}{9 \times 3} \times \frac{4}{4}$. Both methods will leave you with $\frac{14}{27}$.

Canceling allows you to divide either numerator and either denominator by the same number. In that last example, it would look like this:

You need to multiply $\frac{8}{9} \times \frac{7}{12}$. You notice that both the numerator 8 and the denominator 12 have a factor of 4.

Divide each of them by 4, and show that by crossing out the original number and writing the quotient of that number and 4.

Multiply numerator times numerator and denominator times denominator, using the new numbers.

$$\frac{8}{9}\times\frac{7}{12}=\frac{\overset{2}{\cancel{8}}}{9}\times\frac{7}{\cancel{12}_{3}}=\frac{2\times7}{9\times3}=\frac{14}{27}$$

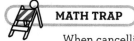 **MATH TRAP**

When cancelling, remember to divide a numerator and a denominator by the same number. Never divide two numerators or two denominators.

Let's look at some examples. First, we'll walk through $\frac{3}{8}\times\frac{4}{9}$. The basic rules says to multiply 3 times 4 and 8 times 9, giving the fraction $\frac{12}{72}$, which reduces to $\frac{1}{6}$. However, you can cancel before multiplying. Divide 3 into both 3 and 9, and divide 4 into both 4 and 8. The problem turns into $\frac{1}{2}\times\frac{1}{3}$.

$$\frac{3}{8}\times\frac{4}{9}=\frac{\overset{1}{\cancel{3}}}{8}\times\frac{4}{\cancel{9}_{3}}=\frac{\overset{1}{\cancel{3}}}{_{2}\cancel{8}}\times\frac{\overset{1}{\cancel{4}}}{\cancel{9}_{3}}=\frac{1\times1}{2\times3}=\frac{1}{6}$$

Let's try one with a mixed number. That's a whole number plus a fraction, but you can turn it into an improper fraction by finding the proper disguise for the whole number part.

Divide $1\frac{1}{14}\div\frac{3}{7}$.

Whenever a multiplication or division problem involves a mixed number, convert to an improper fraction immediately. The 1 is equal to $\frac{14}{14}$ and the additional $\frac{1}{14}$ makes $\frac{15}{14}$, so your problem is really $\frac{15}{14}\div\frac{3}{7}$.

To divide, you need to multiply by the reciprocal of the divisor, or invert the divisor, $\frac{3}{7}$, and multiply. The problem $\frac{15}{14}\div\frac{3}{7}$ is equivalent to $\frac{15}{14}\times\frac{7}{3}$. Cancel before you multiply. Divide both 15 and 3 by 3, and divide 7 and 14 by 7.

$$\frac{15}{14}\times\frac{7}{3}=\frac{\overset{5}{\cancel{15}}}{14}\times\frac{7}{\cancel{3}_{1}}=\frac{\overset{5}{\cancel{15}}}{_{2}\cancel{14}}\times\frac{\overset{1}{\cancel{7}}}{\cancel{3}_{1}}=\frac{5\times1}{2\times1}=\frac{5}{2}$$

The product $\frac{5}{2}$ is an improper fraction, and that's fine. If you'd rather see it as a mixed number, remember the denominator says the whole was broken into two pieces, broken in halves. You have 5 of those, so you can piece together 2 wholes and have 1 half left.

$$\frac{5}{2}=2+\frac{1}{2}=2\frac{1}{2}$$

> **WORLDLY WISDOM**
>
> To quickly convert a mixed number to an improper fraction, multiply the denominator by the whole number and add to the numerator. $6\frac{5}{8} = \frac{8+6+5}{8} = \frac{48+5}{8} = \frac{53}{8}$. To change an improper fraction to a mixed number, divide numerator by denominator and make any remainder the numerator of the fraction. $\frac{43}{7} = 6$ with 1 left over, so $6\frac{1}{7}$.

> **CHECK POINT**
>
> Perform each multiplication or division problem.
>
> 11. $\frac{2}{3} \times \frac{21}{50}$
>
> 12. $\frac{8}{15} \div \frac{2}{25}$
>
> 13. $\frac{6}{49} \times \frac{14}{15}$
>
> 14. $4\frac{3}{8} \div 1\frac{7}{8}$
>
> 15. $5\frac{4}{7} \times 4\frac{2}{3}$

## Adding and Subtracting Fractions

Adding and subtracting fractions requires that the fractions have the same denominator. Denominators tell you what kind of things you have, and it's difficult to even think about adding unless you have the same kind of things, or a *common denominator*. When you have a common denominator, you simply add the numerators and keep the denominator. $\frac{3}{11} + \frac{7}{11} = \frac{10}{11}$. Three of these things plus 7 of these things equals 10 of these same things.

> **DEFINITION**
>
> A **common denominator** is one that is a multiple of each of the denominators of two or more fractions. The least common denominator is the least common multiple of the denominators.

Your real work comes when the fractions don't have a common denominator to start out. It's not impossible to add or subtract fractions with different denominators, but you must change their appearance first. You have to change them to equivalent fractions with a common denominator.

A common denominator is a number that is a multiple of each of the denominators you were given. Ideally, you should choose the lowest number that can be evenly divided by each of the denominators in your equation. That's called the lowest or least common denominator. Larger multiples will work, but you will have to simplify at the end. Let's look at some examples.

Suppose you want to subtract $\frac{1}{3} - \frac{2}{7}$. In order to do this operation, you need to make the denominators of the two fractions match. First, find the lowest common multiple of the two denominators.

The lowest common multiple of 3 and 7 is 21. Multiply $\frac{1}{3} \times \frac{7}{7}$ and $\frac{2}{7} \times \frac{3}{3}$ to change them to fractions with the same value but a different look.

Then you are able to subtract.

$$\frac{1}{3} - \frac{2}{7} = \frac{7}{21} - \frac{6}{21} = \frac{1}{21}$$

Sometimes you'll look at two denominators and know quickly what number can serve as a common denominator, but if the numbers are large, you may need more of a game plan. To find the lowest common denominator, take a moment first to factor each denominator.

Suppose you need to add $\frac{7}{30} + \frac{5}{42}$. You look at 30 and 42 and don't see a common denominator immediately. Think about the factors of each denominator. $30 = 2 \times 3 \times 5$ and $42 = 2 \times 3 \times 7$. Both denominators have factors of 2 and 3, but 30 also has a factor of 5. 42 has a factor of 7. The lowest common denominator will be the product of the 2 and 3 they have in common and the 5 and 7 they don't.

LCD = $2 \times 3 \times 5 \times 7 = 210$

Multiply $\frac{7}{30} \div \frac{7}{7}$, because 30 doesn't have the factor of 7, and $\frac{5}{42} \times \frac{5}{5}$, because 42 doesn't have the factor of 5. Give each fraction what it's missing, then add and simplify.

$$\frac{49}{210} + \frac{25}{210} = \frac{74}{210} = \frac{37}{105}$$

> **WORLDLY WISDOM**
>
> It is not always necessary to have the lowest common denominator. Sometimes you simply want to add or subtract the fractions as quickly as possible. You can fall back on this strategy. Multiply the two denominators for a common denominator. Multiply the lower right denominator times the upper left numerator and put the result in the upper left numerator's position. Multiply the lower left denominator by the upper right numerator and put the result in the upper right numerator's position. Add or subtract the numerators and simplify if necessary.

You can handle mixed numbers in addition and subtraction by changing them to improper fractions, as you did in multiplication, but you don't always have to. In addition, you can just combine the fractions and add the whole numbers on to that sum at the end. $4\frac{3}{5} + 2\frac{5}{8}$ is really $4 + \frac{3}{5} + 2 + \frac{5}{8}$, so you can rearrange to $\frac{3}{5} + \frac{5}{8} + 4 + 2$.

For subtraction, it can be trickier, because there may be regrouping, or borrowing, involved. Try to subtract the fraction parts, and if you can, then you can also just subtract the whole numbers.

$$8\frac{4}{5} - 2\frac{1}{5} = 8 + \frac{4}{5} - \left(2 + \frac{1}{5}\right) = (8-2) + \left(\frac{4}{5} - \frac{1}{5}\right) = 6\frac{3}{5}$$

If you try subtracting the fractions and realize the first fraction is smaller than the second, you need to borrow from the first whole number. To subtract $8\frac{1}{5} - 2\frac{4}{5}$, you need to borrow 1 from the 8.

$$8\frac{1}{5} - 2\frac{4}{5} = 7 + 1 + \frac{1}{5} - 2\frac{4}{5} = 7 + \frac{5}{5} + \frac{1}{5} - 2\frac{4}{5} = 7\frac{6}{5} - 2\frac{4}{5} = 5\frac{2}{5}$$

---

**CHECK POINT**

Complete each addition or subtraction problem.

16. $\dfrac{4}{7} + \dfrac{3}{5}$

17. $\dfrac{8}{9} - \dfrac{1}{4}$

18. $\dfrac{2}{15} + 3\dfrac{1}{8}$

19. $7\dfrac{2}{3} + 9\dfrac{3}{4}$

20. $5\dfrac{1}{10} - 1\dfrac{1}{2}$

## The Least You Need to Know

- To change a fraction to an equivalent form, multiply (or divide) both the numerator and the denominator by the same number.
- To multiply fractions, multiply the numerators, multiply the denominators, and simplify if possible. You may be able to simplify before multiplying by dividing a numerator and a denominator by the same number.
- To divide fractions, invert the divisor and multiply.
- The lowest common denominator of two fractions is the LCM of the two denominators.
- To add or subtract fractions, change each fraction to an equivalent fraction with a common denominator, and add or subtract the numerators. Simplify if possible.

# CHAPTER 6

# Decimals

Not every number system in history had a way to represent parts of a whole but our system does. In fact, it has two. You've already met common fractions, the ones that are written as a quotient of two integers. In this chapter, you'll get to know decimal fractions.

A decimal fraction, or as most people call it, a decimal, is a representation of part of a whole written in a way that fits into our decimal place value system. In this chapter, you'll see how to extend the place value system to include parts of a whole, and how to write very small numbers in scientific notation. You'll look at each of the operations of arithmetic when decimals are involved, and you'll see that one of the advantages of decimals is that they fit nicely into the methods you already know. Just in case you need to switch between the two ways of writing fractions, you'll learn to convert from common fractions to decimal fractions and back.

## In This Chapter

- The system of decimal fractions
- Writing very small numbers in scientific notation
- Adding and subtracting decimals
- Multiplying and dividing decimals
- Converting between fractions and decimals

# Decimal Fractions

By now you're familiar with the fractions that are called common, the ones written as a quotient. Now it's time for a look at decimal fractions. They're fractions that only use denominators that will fit into the decimal, or base ten, system. Decimal fractions have denominators of 10, or 100, or 1000 and so on, but you hear the denominator in the name rather than seeing it under a fraction bar. The common fraction $\frac{3}{10}$ has a numerator of three and a denominator of 10, so you read it as "three tenths." When that same number is written as a decimal fraction, it's 0.3. It's still read "three tenths," because the 3 is in the tenths place.

The places that follow the decimal point of a number are assigned to be tenths, hundredths, thousandths, and so on, so that the place value tells the denominator. 0.1 is 1 tenth, and 0.01 is 1 hundredth. What would have been the numerator of the fraction becomes the digit in the place that corresponds to its denominator. 0.7 is 7 tenths, the equivalent of $\frac{7}{10}$, 0.008 is 8 thousands, or $\frac{8}{1000}$, and 0.91 is 91 hundredths, or $\frac{91}{100}$. When you write a fraction in the decimal system, the denominator seems to disappear, but you hear it if you give the decimal its proper name.

**MATH TRAP**

While many people will say "point-3-7-5" for 0.375, its real name is "three hundred seventy-five thousandths." The "three hundred seventy-five" is the numerator and the "thousandths" names the denominator.

Place a dot, called a decimal point, to the right of the ones digit of a number. This is the divider between the whole and the parts. You've probably seen this idea used to write dollars and cents. Three dollars and 12 cents might be written $3.12. This says 3 whole dollars (left side of the decimal point) and 12 pennies, part of a dollar (right side of the decimal point).

## Numbers Less than One

Our decimal system is able to represent cents as part of a dollar, but also to represent even smaller parts of any whole, again by working with tens. Decimal fractions, as their name says, are still based on tens. On the left side of the decimal point, as you move to the left, away from the decimal point, the value of that place gets bigger. Specifically, it gets ten times larger. On the right side of the decimal point, each time you move to the right away from the decimal point, the place value will get smaller. With each step, you'll break the piece into ten smaller pieces.

The decimal fractions are written to the right of the decimal point. The first position after the point is the tenths place. It imagines you broke a whole into ten parts, called tenths, and the digit in that place shows how many of those parts you have. In our dollar and cents example, the 1 just to the right of the decimal point is worth one tenth of a dollar, or in other words, a dime.

How many tenths can you have? You might not have any, in which case you could put a zero in that place. If you had 1, or 2, or 3, or 4, or 5, or 6, or 7, or 8, or 9, that digit goes in the tenths place. What if you have more than that? Ten tenths make 1 whole, so make groups of ten tenths, see how many groups you have and put that number in the ones place. The leftovers, if any, will have to be fewer than 10 tenths, so that digit goes in the tenths place. If you had 23 tenths, you'd have 2 groups of 10 tenths and 3 tenths left over, so that would be 2 ones and 3 tenths, or 2.3.

Of course, you might want to break the whole into smaller parts, so the next place to the right imagines you break the whole into 100 parts, each called one hundredth. You can think of it as breaking each tenth into ten parts, or breaking one whole into a hundred parts. You get the same thing.

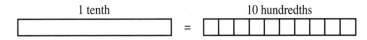

The digit in the hundredths place tells you how many of those little pieces you have. The number 4.05 says you have 4 ones and 5 hundredths.

> 💡 **WORLDLY WISDOM**
>
> Notice that all the places to the right of the decimal have a "th" at the end of their names. That's the signal that you're talking about a fraction.

Do you have more than 9 of these hundredths? A group of 10 hundredths makes one tenth, so count up how many groups you have and that goes in the tenths place, and what's left over goes in the hundredths place. You can think of the number 0.48 as 48 hundredths, or as 4 tenths and 8 hundredths.

If you had more than a hundred hundredths, you'd make a 1 from each group of 100. When you've made all the groups of 100 you can, you can look for groups of 10 hundredths to make tenths, and the leftovers would go in the hundredths place. If you had 793 hundredths, you'd have 7 groups of 100 hundredths, 9 groups of 10 hundredths, and 3 hundredths left over. That's 7 ones, 9 tenths and 3 hundredths, or 7.93, seven and ninety-three hundredths.

# Different Names, Same Number

Decimal fractions can sometimes go by more than one name. If you write 3.82, you read that as "three and eighty-two hundredths." You have three wholes and eighty-two of the fractions that are created by breaking one into 100 parts. Those eighty-two hundreds can make 8 groups of ten hundredths or eight tenths, with two hundredths left over, so there's an 8 in the tenths place and a 2 in the hundredths place.

Now suppose you write 3.820, which you'd read as "three and eight hundred twenty thousandths." You read this number differently because it fills in the thousandths place, but its value is exactly the same as 3.82. All you added was zero thousandths—nothing!

Trailing zeros at the end of a decimal fraction change the way you read the name of the number, but they don't change the value of the number. The numbers 1.7, 1.70, and 1.700 all have the same value. They would be read as "one and seven tenths," "one and seventy hundredths," and "one and seven hundred-thousandths," but they're all worth the same amount. If you wrote them as common fractions, 1.7 would be $1\frac{7}{10}$, 1.70 would be $1\frac{70}{100}$, which reduces to $1\frac{7}{10}$, and 1.7000 would be $1\frac{700}{1000}$, which also reduces to $1\frac{7}{10}$. As a result, you'll often want to drop any trailing zeros to make the look of the number simpler.

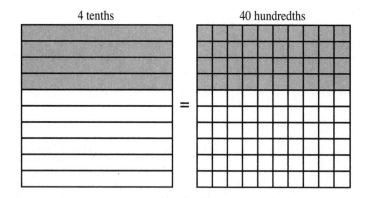

4 tenths         =         40 hundredths

1. Write the number 9.003 in words.

2. Write the number 82.4109 in words.

3. Write the number "forty-two hundredths" in numerals.

4. Write the number "forty-two ten-thousandths" in numerals.

5. Write the number "three hundred twelve and nine hundred one thousandths" in numerals.

# Powers and Scientific Notation Revisited

Earlier you learned about using exponents to write powers of ten and how to write large numbers as the product of a number between 1 and 10 and a power of 10. What about the other side of the decimal point? How can you write the values of the places that hold decimal fractions as powers of ten? And how do you use scientific notation to express tiny decimal fractions?

## Powers of Ten for Numbers Less than 1

Our place values to the left of the decimal point are $10^0$, $10^1$, $10^2$, and so on, with the exponent getting larger as you move to the left. Each decimal fraction was created by breaking a whole into ten pieces or a hundred pieces or a thousand pieces and so on. You're still using powers of 10, but you're dividing by 10 each time you move to the right. You're looking at place values that are getting smaller instead of bigger.

To write powers of 10 for these fractional values, you're going to use a very similar system of exponents, but you have to add a signal that you're going to the other side of the decimal point, going in the opposite direction. The signal that you're moving to the right of the decimal point is a negative exponent. The negative signals an opposite, in this case, a move in the opposite direction.

The tenths place is worth a tenth of a whole, or the fraction you get by dividing 1 by 10. You signal that value as $10^{-1}$. The hundredths place is worth $1 \div 100$, or $1 \div 10^2$, which you can write as $10^{-2}$. The thousandths place is worth $10^{-3}$. To express the value of a place to the right of the decimal point as a power of ten, count the number of digits to the right of the decimal point, put a minus sign in front of that number, and use that number as an exponent on 10.

### Numbers Less than 1 as Powers of 10

| Tenths | Hundredths | Thousandths | Ten-thousandths | Hundred-thousands |
|---|---|---|---|---|
| 0.1 | 0.01 | 0.001 | 0.0001 | 0.00001 |
| 1 digit after decimal point | 2 digits after decimal point | 3 digits after decimal point | 4 digits after decimal point | 5 digits after decimal point |
| $10^{-1}$ | $10^{-2}$ | $10^{-3}$ | $10^{-4}$ | $10^{-5}$ |

**CHECK POINT**

6. Write 0.00001 as a power of ten.

7. Write 0.000000001 as a power of ten.

8. Write 0.000000000000001 as a power of ten.

9. Write $10^{-6}$ in standard form.

10. Write $10^{-10}$ in standard form.

# Scientific Notation for Small Numbers

Writing small numbers in scientific notation is almost the same process as you used for large numbers. A small number like 0.00045 can be written in scientific notation by first writing the digits of the number and placing a decimal point after the first nonzero digit. This gives you the number between 1 and 10. (Later you can drop any leading zeros.)

Count from where you just placed the decimal point to where it actually should be. Notice that you're counting in the opposite direction from what you did with large numbers. You'll show that by making the exponent negative. The number of places tells you the exponent to put on the ten, but it will have a minus sign on it.

Suppose you want to write 0.00045 in scientific notation.

Copy the number and put a decimal point after the first nonzero digit.

0 . 0004 . 5
actual            new
decimal          decimal
point             point

Count to where the decimal point ought to be.

0.0004.5
  4 places

Make the exponent negative.

00004.5 needs an exponent of -4.
4 places

Write as a number between 1 and 10 times a (negative) power of ten.

$00004.5 \times 10^{-4}$

Drop any leading zeros.

$4.5 \times 10^{-4}$

Now you can see that $0.00045 = 4.5 \times 10^{-4}$.

> **MATH TRAP**
>
> It's easy to get confused with so many zeros in the number. Count carefully, and place a mark over or under each digit as you count your way to the decimal point. Have a system so you don't get confused with all those zeros.

To change a small number that is written in scientific notation to standard form, copy the digits of the number between 1 and 10 and move the decimal point to the left as many places as the exponent on the 10. This is the same process you used with large numbers, but you move to the left because, for small numbers, the exponent is negative. You can add zeros if you run out of digits. The number $4.193 \times 10^{-4}$ becomes $\underset{\text{4 left}}{0004}.193$ or .0004193. It's customary to put a zero in the ones place, so this could be written 0.0004193.

**CHECK POINT**

11. Write 0.492 in scientific notation.

12. Write 0.0000051 in scientific notation.

13. Write $2.7 \times 10^{-5}$ in standard notation.

14. Write $8.19 \times 10^{-7}$ in standard notation.

15. Write $5.302 \times 10^{-4}$ in standard notation.

# Arithmetic with Decimal Fractions

Just about any work you can do with whole numbers or integers, you can do with fractions, whether they're written as common fractions or as decimals. Because decimal fractions fit into the base ten system, many of the operations with decimals are the same as, or very similar to, the operations with whole numbers.

## Adding and Subtracting Decimals

You've already seen that decimal fractions can be written in the same place value system we use for whole numbers. When you add or subtract whole numbers, you align the right end of the numbers. You're actually aligning the unseen decimal points, so that digits with the same place value are under one another.

To add or subtract decimals, first arrange the numbers with the decimal points aligned one under another. This assures that you are adding the digits with the same place value. If the numbers don't have the same number of digits after the decimal point, you can add zeros to the shorter ones. Remember that adding a trailing zero changes the way you name a number, but not what it's worth.

Add or subtract as you would if the decimal points were not there, starting from the right, doing any carrying or borrowing just as you would for whole numbers. Add a decimal point to your answer directly under the decimal points in the problem.

For example, here's how you add 34.82 and 9.7.

Align the decimal points and add a zero to 9.7.

```
  1 1
 34.82
  9.70
 44.52
```

Add as you normally would. The hundredths place adds to 2 hundredths, so you can just put that down, but the tenths place adds to 15 tenths, so you need to carry 1 to the ones place. You'll carry again from ones to tens. You work right over the decimal points as though they weren't there. Bring the decimal point straight down when you're all done.

Subtraction works the same way. Align the decimal points, and add zeros to make the decimals the same length. Those zeros are especially important in subtraction, because you can't subtract from a digit that isn't there. Subtract normally, regrouping if you need to, and bring the decimal point straight down. Let's try a subtraction example.

Subtract 184.93 − 99.781

Align the decimal points and add a trailing zero to 184.93. Subtract, starting at the right. Regroup or borrow as you need to (and you need to a lot for this one). Bring the decimal point straight down.

```
   7    8  ¹2
 18 ¹4.9 3 ¹0
 ‾
    9 9.7 8 1
   8 5.1 4 9
```

You can check your answer to a subtraction problem by adding. In the example above, check to see if 85.149 + 99.781 gives you 184.930. If it does, your subtraction is correct.

> **WORLDLY WISDOM**
>
> Taking a minute to estimate you answer before you do a calculation is always a good idea, and with decimals, it will help you be certain you got the decimal point in the right place. It doesn't need to be anything fancy. If you can say that you're subtracting about 185 - 100, you'll know you should be getting an answer around 85. If you get an answer more like 8 or 850, you've misplaced the decimal point.

---

## Multiplying and Dividing Decimals

Multiplying decimals requires a slightly more complicated process. It's a good idea to estimate the product first, so if you have to multiply $3.1 \times 2$, you'd expect an answer around $3 \times 2$ or 6. If you multiply $3.1 \times 0.002$, you'd recognize that you're multiplying a number around 3 by a pretty small number, quite a bit less than 1, so the answer should be less than 3, probably a lot less.

You perform the actual multiplication as though the decimal points were not present. If you were multiplying 3.1 times 2, you would first think of it as 31 times 2. (You would also think of 3.1 times 0.002 as 31 times 2 to begin. The difference comes in placing the decimal point.)

To place the decimal point in the product, first count the number of digits to the right of the decimal point in each of the factors being multiplied. Add these counts up to find the number of decimal places in the product. Start from the far right end of the answer and count to the left to place your decimal point.

For example, multiply $3.1 \times 2$.

The first factor, 3.1, has one digit after the decimal point. The second factor, 2, has none, so the product will have one digit after the decimal point.

Once you know that, multiply as if the decimal point wasn't there. $31 \times 2 = 62$. Then place the decimal point so that your answer has one digit after the decimal point. $3.1 \times 2 = 6.2$.

You estimated the answer as approximately 6, so 6.2 sounds quite reasonable. Let's look at one where the multiplier is smaller.

Multiply $3.1 \times 0.002$.

This problem has the same important digits, so you'll still start by thinking that $31 \times 2 = 62$, but the decimal points are positioned differently.

Count the decimal places. There is one digit after the decimal point in 3.1. There are three digits after the decimal point in 0.002, so the answer must have a total of 4 digits after the decimal point.

Start from the right side and count to the left. The 2 in 62 is one digit, the 6 is a second, but there must be 4 digits to the right of the decimal point so place zeros for the third and fourth, giving an answer of 3.1 × .002 = .0062.

Once again, it's traditional to put a zero in the ones place, so 3.1 × 0.002 = 0.0062.

> **WORLDLY WISDOM**
>
> Just as you can add trailing zeros at the right end of a decimal without changing its value, you can add zeros at the left of a number. 87 and 087 describe the same number, but with whole numbers you don't need that zero in front. When you're placing a decimal point you might. If you run out of digits, add zeros to the left end of the number until you have enough places.

To multiply decimals:

- Multiply the factors without regard to the decimal points.

- Count the number of digits that follow the decimal point in each factor.

- Add those counts to find the number of digits after the decimal point in the product.

- Start from the right end of the product and count to the left until there are the required number of digits after the decimal point.

Just as multiplication of decimals is based on multiplication of whole numbers, with a rule for placing the decimal point, the division of decimals is based on whole number long division. But again, there will be some extra rules about decimal points. Let's sneak up on it a little at a time.

To divide a decimal by a whole number, do your division as though both numbers were whole numbers, but then let the decimal point rise straight up into the quotient. Suppose you want to divide 54.96 ÷ 12. You're going to divide a number close to 55 by 12. 12 × 4 = 48, and 12 × 5 = 60, so your quotient should be between 4 and 5.

Divide as if you were dividing 5,496 by 12, and for now, just let the decimal point sit between the 4 and the 9. When you're done dividing, let the decimal point float straight up, and it will land right after the 4 in the quotient, giving you an answer of 4.58.

```
        4.58
       ⎯⎯⎯⎯⎯
       ↑
   12⟌54.96
       48
       ⎯⎯
        69
        60
        ⎯⎯
         96
         96
         ⎯⎯
```

So if that's how you divide a decimal by a whole number, how do you divide by a decimal? You don't. What you do is actually to find an equivalent problem, a problem with the same answer, in which the divisor is a whole number.

That may sound like magic, but if you stop to think about it, you've seen equivalent problems. You know that $12 \div 4 = 3$ and $120 \div 40 = 3$. Those two different problems have the same answer because both the dividend and the divisor were multiplied by the same number, in that case, by 10. That will be the secret to decimal division.

 **MATH TRAP**

When you divide decimals, the divisor needs to be a whole number. The dividend doesn't. Move the decimal point as many places as it takes to turn the divisor into a whole number. Move the decimal point in the dividend the same number of places, whether it becomes a whole number, or still has decimal digits, or needs trailing zeros added so that you can move enough. The divisor calls the play. Whatever happens to the divisor happens to the dividend.

When you multiply a decimal by 10, the decimal point moves 1 place to the right. The task of dividing 81.312 by 3.52 can also be thought of as 813.12 divided by 35.2, or 8131.2 divided by 352.

When you're faced with a division problem where the divisor is a decimal, move the decimal point in the divisor to the right until the divisor is a whole number. Move the decimal point in the dividend the same number of places to the right. Divide normally, just as you would for whole numbers, and bring the decimal point straight up into the quotient. Here's an example.

Divide 17.835 by 2.05.

Estimate first. A number almost 18 divided by about 2 should give you an answer close to 9.

Move the decimal point in the divisor, 2.05, two places right so that 2.05 becomes 205. Move the decimal point in 17.835 two places right as well, making it 1783.5. Then divide.

$$
\begin{array}{r}
8.7 \\
205 \overline{)1783.5} \\
\underline{1640} \\
1435 \\
\underline{1435} \\
0
\end{array}
$$

The decimal point in the dividend, between the 3 and the 5, floats straight up into the quotient for a result of 8.7.

> **WORLDLY WISDOM**
>
> If you've done your division and still have a remainder, you can add zeros to the dividend and keep dividing. If your dividend was a whole number and you still have a remainder when you've divided through to the ones place, add a decimal point and some zeros. Then keep dividing to get a decimal in the quotient.

Step-by-step, the process for long division of decimals looks like this:

- Make an estimate of the quotient.

- Move the decimal point to the end of the divisor.

- Move the decimal point an equal number of places in the dividend.

- Divide normally and let the decimal rise directly up into the quotient.

> **CHECK POINT**
>
> Complete each multiplication or division problem.
>
> 21.  $4.92 \times 1.5$             24.  $461.44 \div 1.12$
>
> 22.  $68.413 \times 0.15$          25.  $5,066.518 \div 8.6$
>
> 23.  $95.94 \div 7.8$

## Converting Fractions

To convert a common fraction to a decimal, do what the fraction tells you. No, they don't actually talk, but every fraction is a division problem. The fraction $\frac{3}{4}$ is the number you get when 3 is divided by 4. To change it to a decimal, do the division.

> **WORLDLY WISDOM**
>
> You can use the word TIBO to remind you that when you convert a fraction to a decimal, you put the top number, or numerator, inside the division sign, and the bottom number, or denominator, outside and then divide. Top In, Bottom Out, or TIBO.

To change $\frac{5}{8}$ to a decimal, do the division problem $8\overline{)5}$.

Add a decimal point after the 5 and add some zeros. Do the division, letting the decimal point from the dividend float straight up to the quotient.

$$
\begin{array}{r}
0.625 \\
8\overline{)5.000} \\
\underline{48} \phantom{00} \\
20 \phantom{0} \\
\underline{16} \phantom{0} \\
40 \\
\underline{40}
\end{array}
$$

If you're changing a fraction to a decimal and you find that you've gotten several digits in the decimal form and you're not getting a zero remainder, you're not expected to keep dividing forever. You might carry the division one place farther than you actually want and then round your answer, understanding that this will give you an approximate, not exact, representation of the fraction.

$\frac{4}{7}$, for example, converts to 0.571428… and keeps going. You might round it to 0.571. On the other hand, you might notice that as you divide, a pattern emerges. $\frac{7}{9}$ converts to 0.77777… and keeps repeating. To show that a pattern keeps repeating, you put a bar over the top, like this: $\frac{7}{9} = 0.\overline{7}$.

To convert a decimal to a fraction, many times all you have to do is say the decimal's name. The decimal 0.37 is "37 hundredths," so write it as $\frac{37}{100}$. If possible, simplify the fraction. The decimal 0.8 is "eight tenths," or $\frac{8}{10}$. That simplifies to $\frac{4}{5}$.

Repeating decimals are a little trickier to convert to fractions. If you need to convert a repeating decimal like 0.12121212… or $0.\overline{12}$ to a fraction, make the pattern (in this case, 12) your numerator, and then count the number of digits in the pattern (in this case, 2 digits). Make your denominator that many nines. $0.\overline{12}$ is equal to 12 over 2 nines or $0.\overline{12} = \frac{12}{99}$. Of course, simplify if you can. $0.\overline{12} = \frac{12}{99} = \frac{4}{33}$.

---

**CHECK POINT**

26. Change to a decimal: $\frac{4}{25}$

27. Change to a decimal: $\frac{7}{9}$

28. Change to a fraction: 0.185

29. Change to a fraction: $0.\overline{123}$

30. Change to a decimal: $\frac{59}{8}$

# Rational and Irrational Numbers

When you met rational numbers, you might have wondered if there were any numbers that aren't rational, and there are. They're not whole numbers, but you can't find a way to write them as common fractions. It's difficult to write an irrational number, because it can't be written as a fraction, and when you try to write it as a decimal, it goes on forever. There are rational numbers whose decimal form goes on forever, but those are repeating decimals, like $\frac{2}{3} = 0.66666...$, which you can write as $0.\overline{6}$. The irrational numbers don't have repeating patterns.

Taken together, the rational numbers and the irrational numbers form the set of *real numbers*.

 **DEFINITION**

**Irrational numbers** are numbers that cannot be written as the quotient of two integers. **Real numbers** is the name given to the set of all rational numbers and all irrational numbers.

Some of the irrational numbers come up when you try to take the square root of a number that isn't a perfect square number. Taking the square root of a number means finding some number you can multiply by itself to give you this answer. The square root of 9 is 3, because $3^2 = 9$. We write $\sqrt{9} = 3$.

If one number can be squared to produce another, the first number is the square root of the second. The square root of 16 is 4 because $4^2 = 16$. The symbol for square root, $\sqrt{\phantom{x}}$, is called a radical. $\sqrt{16} = 4$, $\sqrt{6.25} = 2.5$, and $\sqrt{2}$ is an irrational number approximately equal to 1.414.

When you try that with a number that isn't a square number, like 8, the answer isn't an integer, and many times, it's an irrational number. $\sqrt{8} = 2.828427125$. When that happens, you either have to round that number and say, for example, that $\sqrt{8}$ is approximately 2.83 or just leave the number in square root notation: $\sqrt{8}$.

Many times your first encounter with an irrational number comes when you work with circles and meet the number called *pi*, or $\pi$. The reason it has a name is that it is difficult to write. It's a number a little larger than 3, but there's no fraction that's exactly right, and its decimal form goes on forever without a pattern. People have used approximate values like $\frac{22}{7}$ or 3.14, and those come close, but they're not exactly the number we call pi. Mathematicians have worked at finding many of the decimal digits of pi. We can say that pi is approximately 3.14159265359… but it just keeps going, and as you can see, there's no repeating pattern.

Those are the characteristics of an irrational number. It can't be written as a fraction, because it's not rational. When you try to find a decimal expression for it, it goes on and on and doesn't show you any pattern.

**MATH IN THE PAST**

*Pi* is the name given to the ratio of a circle's circumference, the measurement around the circle, to its diameter, the measurement across the widest part. *Pi* has a long history. There is evidence that mathematicians were thinking about this number in ancient Babylon, Egypt, Israel, Greece, China, and India, but all of them came up with different values a bit more than 3. Today, mathematicians have calculated pi to trillions of digits. Still no pattern.

The rational numbers and the irrational numbers together make up the real numbers. The real numbers include all the other sets of numbers we've looked at. We started with the counting numbers and then got more complex. The irrational numbers are the first set that doesn't overlap the ones you've met before.

You might picture the set of real numbers something like this:

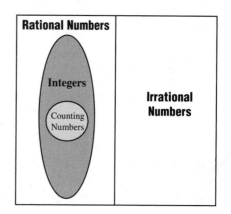

**WORLDLY WISDOM**

Are there fake numbers? If the rational numbers and the irrational numbers combine to make up the real numbers, are there numbers that aren't real numbers? Actually, mathematicians will tell you that there are, but they don't call them fake. They call them imaginary numbers, because they aren't real numbers, but you can imagine them. You know that $2^2 = 4$, and it's also true that $(-2)^2 = 4$, but what number can you square to get -4? Most people would say there is none, but if you let yourself imagine it, there can be a $\sqrt{-4}$.

## The Least You Need to Know

- The decimal point separates whole numbers on the left from fractions on the right. As you move to the right, the value of each place is divided by 10.

- To write a small number in scientific notation, write it as a number greater than or equal to 1 and less than 10 times a negative power of 10.

- Performing arithmetic operations with decimals is very similar to performing arithmetic operations with whole numbers.

- To do arithmetic with decimal fractions, work as though they were whole numbers, then follow the rules for placing the decimal point.

- To change a common fraction to a decimal, divide the numerator by the denominator.

- To change a terminating decimal to a fraction, put its digits over the appropriate power of 10. To change a repeating decimal to a fraction, put the digits in the repeating pattern over the same number of 9s.

# Ratios, Proportions, and Percentages

We have two basic ways of comparing numbers: an addition/subtraction method and a multiplication/division method. The addition/subtraction method makes statements like 15 is 7 more than 8, or 49 is 1 less than 50. The multiplication/division method compares numbers by saying things like 63 is 3 times as large as 21, or 12 is half of 24.

In this chapter, you'll focus on that comparison by multiplication and division. You'll learn to use ratios and extended ratios to help you figure out unknown numbers, and you'll solve proportions by cross-multiplying. Once you understand percentages, you'll be able to explore how they're using in different kinds of problem solving

## In This Chapter

- How to compare numbers by creating ratios
- Using ratios to solve problems
- Using proportions to find unknown numbers
- Changing a ratio to a percentage
- How to calculate interest, tax, and tips

## Proportional Reasoning

Twice the size, half as many, three times as much. In your daily language, you frequently use the idea of multiplying or dividing as a way to compare numbers. Usually, in conversation, you'll make the comparison using a simple whole number compared to 1: twice the size or three times as much. Or you might use division in the form of a simple fraction, like half as many. But you could make other comparisons, not always comparing to 1.

If you had two lamps, one 2 feet tall and one 3 feet tall, you could say one is 1.5 times taller than the other, but more often, you'll stay with the whole numbers and say the taller lamp is to the shorter one as 3 is to 2. Or you could say the shorter lamp is to the taller one as 2 is to 3.

## Understanding Ratios and Extended Ratios

A *ratio* is a comparison of two numbers by division. If one number is three times the size of another, you say the ratio of the larger to the smaller is "3 to 1." This can be written as 3:1 or as the fraction $\frac{3}{1}$. You could also compare the smaller to the larger by saying the ratio is 1 to 3.

**DEFINITION**

A **ratio** is a comparison of two numbers by division.

When you are told that the ratio of one number to another is 5:2, you are not being told that the numbers are 5 and 2, but that when you divide the first by the second, you get a number equal to $\frac{5}{2}$. This happens when the first number is 5 times some number and the second is 2 times that number. If that's all you know about the numbers, you have a sense of their relative size, but that's about all. If you have some other information, you might be able to figure out what the numbers actually are.

Suppose two numbers are in ratio 7:3 and their sum is 50. You know that the first of the numbers you're looking for is 7 times some number and the other is 3 times that same number, so that they divide to $\frac{7}{3}$. With the extra piece of information that they add to 50, you can try to find the numbers by guess and test. If the numbers were actually 7 and 3, they would add to 10. Multiply each one by 2 and you have 14 and 6, which add to 20. A little more experimentation will tell you that 7 × 5, or 35, and 3 × 5, or 15, will add to 50.

That guess and test method can run into some problems. For example, most people won't guess anything but whole numbers, and the answer won't always be a whole number, or it might be a very large whole number. Try using a letter, maybe $n$, called a *variable*, for the number you don't know. If the numbers are in ratio 6:5, one is $6n$ and the other is $5n$. If they add to 88, you can say $6n + 5n = 88$. That means that $11n = 88$, and $n$ must be 8. The numbers are 6 × 8 = 48 and 5 × 8 = 40.

**DEFINITION**

A **variable** is a letter or other symbol that takes the place of an unknown number.

An *extended ratio* compares more than two numbers. Extended ratios are usually written with colons, because you don't want to put more than two numbers into a fraction. Extended ratios are actually a condensed version of several ratios. If the apples, oranges, and pears in a fruit bowl are in ratio 8:3:2, it means that that the number of apples is 8 times some number, the number of oranges is 3 times that number, and the number of pears is 2 times that number. It also means that the ratio of apples to oranges is 8:3, the ratio of oranges to pears is 3:2 and the ratio of apples to pears is 8:2.

📖 **DEFINITION**

> An **extended ratio** combines several related ratios into one statement. It is a way to express the ratios a:b, b:c, and a:c in one statement: a:b:c.

Suppose that a smoothie contains pomegranate juice, orange juice, and yogurt, in a ratio of 2:5:3. If you want to make 5 cups of the smoothie to sip throughout the day, how much of each ingredient will you need? If the numbers actually were 2 and 5 and 3, they would add to 10 cups. You need 5 cups, so let's say *2n* cups of pomegranate juice, *5n* cups of orange juice, and *3n* cups of yogurt, to make a total of 5 cups.

$2n + 5n + 3n = 5$

$10n = 5$

$n = \dfrac{1}{2}$

The multiplier is $\dfrac{1}{2}$, so you need $2 \times \dfrac{1}{2} = 1$ cup of pomegranate juice, $5 \times \dfrac{1}{2} = 2\dfrac{1}{2}$ cups of orange juice, and $3 \times \dfrac{1}{2} = 1\dfrac{1}{2}$ cups of yogurt.

🪜 **MATH TRAP**

> When you're working with ratios that involve measurements, make sure the units match. If you try to say the ratio of the length to the width of a room is 15 feet to 120 inches, when you go on to use that relationship, you'll be confused about whether your numbers are feet or inches, and you'll likely get the wrong numbers. Make it 15 feet by 10 feet or 180 inches by 120 inches, and your work will be easier and more accurate.

1. If the ratio of girls to boys in Math Club is 5:3, and there are 32 members of the club, how many boys are members?

2. The local car dealer sold 40 cars last month. The ratio of gas-powered vehicles to hybrids was 7:1. How many hybrids were sold?

3. The desired ratio of the red roses to white roses in a bouquet is 2:3. If the florist wants the bouquet to have a total of 20 roses, how many white roses and how many red roses will she need?

4. If the ratio of lions to tigers to bears at the zoo is 4:7:4, and there are 45 of these animals all together, how many tigers are there?

5. The ratio of red balloons to white balloons to blue balloons in the auditorium is 21:20:9. If there are 900 balloons in the auditorium, how many are blue?

## Solving Proportions

A *proportion* is a statement that two ratios are equal. The equation $\frac{1}{3} = \frac{2}{6}$ is an example of a proportion. It says the ratio of 1 to 3 is the same as the ratio of 2 to 6, or, in other words, that the fractions are equal.

When you look at the statement that two ratios are equal as a proportion, you can talk about the four numbers that make up the proportion as the *means* and the *extremes*. It's a little easier to understand which numbers are means and which are extremes and why if you write the ratios with colons, so let's write $\frac{1}{3} = \frac{2}{6}$ as 1:3 = 2:6. The extremes are the numbers on the ends, the 1 and the 6. They're far out. The word mean talks about the middle, so the numbers in the middle, the 3 and the 2, are the means. When you do write it as a pair of equal fractions, it looks like this.

$$\frac{\text{extreme}}{\text{mean}} = \frac{\text{mean}}{\text{extreme}}$$

📖  **DEFINITION**

A **proportion** is two equal ratios. The **means** of a proportion are the two middle numbers. The **extremes** are the first and last numbers.

Proportions can be used to compare things. When you create a proportion, it's important to be consistent about the order. If you want to say that the ratio of the shortest side of the big triangle to the shortest side of the small triangle is equal to the ratio of their longest sides, be sure that your second ratio is big triangle to small triangle just like the first ratio. If you change the order, the proportion won't be true.

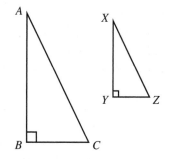

In any proportion, the product of the means—the two middle terms—is equal to the product of the extremes—the first and last terms. For example, in the proportion $\frac{5}{8} = \frac{15}{24}$, $8 \times 15 = 5 \times 24$.

This multiplying of means and extremes is called *cross-multiplying*. Whenever you have two equal ratios you can cross-multiply, and the two product will be equal. Knowing that will often let you find a number that's missing from the proportion. You could use a question mark or other symbol to stand for the missing number, but let's use an $x$ for now.

When you multiply with a variable, especially when you use the variable $x$, it's easy to confuse the variable with the times sign $\times$, so you may want to use other ways to write multiplication, like a dot or parentheses. Instead of writing $5 \times x$, you can write $5 \cdot x$ or $5(x)$.

Keep in mind that cross-multiplying can only be done in a proportion. You can cross-multiply when you have two equal ratios, but not in any multiplication with fractions.

> 📖 **DEFINITION**
>
> Finding the product of the means and the product of the extremes of a proportion, and saying that those products are equal, is called **cross-multiplying**.

Suppose you're told that two numbers are in ratio 7:4, and the smaller number is 14. But what's the larger number? You can use the means-extremes properties of proportions to help you find out.

If I give you the proportion $\frac{x}{14} = \frac{7}{4}$, you can use cross-multiplying to help you find the value of the number I called $x$. If you multiply the means you get 4 times the unknown number or $4x$. The product of the extremes, $7 \times 14$, is 98. The product of the means is equal to the product of the extremes, so $4x$ must equal 98, or $4x = 98$. Dividing 98 by 4 tells you that $x = \frac{98}{4} = 24.5$.

**WORLDLY WISDOM**

When you're cross-multiplying, don't be in a rush to do the arithmetic, especially if your numbers are large. Your work may be easier if you write out the multiplication but don't actually do it. Then when you divide to find the value of the variable, you may see a shortcut. Why do the work of multiplying $42 \times 35$ if, a moment later, you're going to divide by 21? $\dfrac{\overset{2}{\cancel{42}} \cdot 35}{\underset{1}{\cancel{21}}} = 2 \times 35 = 70$. Leave the multiplication in factored form, and the division may be easier.

**CHECK POINT**

Solve the proportion to find the value of the variable.

6. $\dfrac{2}{5} = \dfrac{x}{15}$

7. $\dfrac{3}{7} = \dfrac{24}{x}$

8. $\dfrac{x}{63} = \dfrac{15}{27}$

9. $\dfrac{3}{x} = \dfrac{51}{68}$

10. $\dfrac{3}{5} = \dfrac{x}{10}$

## Percent as a Standard Ratio

The word *percent* means "out of 100." If you shoot 100 free throws at basketball practice and you sink 60 of them, you made 60 out of 100 shots, or 60 percent. But what if you shot 75 and made 40? What percentage is that? Is that better or worse than 60 percent? You can use proportions to convert any ratio to a percentage.

**DEFINITION**

A **percentage** is a ratio that compares numbers to 100. 42 percent means 42 out of 100, or 42:100. Percentages are sometimes written using the percent symbol, %.

Ratios can be hard to compare if they are "out of" different numbers. Which is larger: 4 out of 9 or 5 out of 12? If you compare them as fractions, $\dfrac{4}{9}$ and $\dfrac{5}{12}$, it might help to change to a common denominator. You could use 36 as the common denominator so $\dfrac{4}{9} = \dfrac{16}{36}$ and $\dfrac{5}{12} = \dfrac{15}{36}$, and you can see that $\dfrac{4}{9}$ is larger.

Changing ratios to percentages is like changing fractions to a common denominator. Percentages make it easier to compare different ratios, because they express everything as a part of 100.

When you need to change a ratio to a percentage, take your ratio, like $\frac{15}{20}$, and set it equal to the ratio $\frac{P}{100}$. You're saying 15 is to 20 as $P$ is to 100, or 15 out of 20 is the same as $P$ out of 100. That's a percentage. Use cross-multiplying to find $P$.

$$\frac{15}{20} = \frac{P}{100}$$
$$20P = 1,500$$
$$P = \frac{1,500}{20} = 75$$

So 15 out of 20 is 75 percent.

The rule you want to remember is "part is to whole as percent is to 100" or $\frac{\text{part}}{\text{whole}} = \frac{\%}{100}$. This proportion can be used to solve almost all percent problems. Percentage problems come in three basic types. One type asks a question like "what is 45 percent of 250?" The second type asks "15 is 30 percent of what number?" and the third is "29 is what percentage of 58?" The first type asks you to find the part, the second asks you to find the whole, and the last asks for the percentage.

Suppose your town requires that a candidate receive at least 51% of the vote to be elected mayor. If the population of the town is 1,288, what is the minimum number of votes a candidate must receive to be elected?

You know the percentage and the whole, and you're looking for the part of the population that is required for election. Start with $\frac{\text{part}}{\text{whole}} = \frac{\%}{100}$ or $\frac{p}{1288} = \frac{51}{100}$. Cross-multiply, and you'll find that the product of the means is $1,288 \times 51 = 65,688$ and the product of the extremes is $100p$. The product of the means is equal to the product of the extremes, so $100p = 65688$, and dividing by 100 tells you $p = 656.88$. (Since no one can cast a fraction of a vote, round this to 657 votes.)

**WORLDLY WISDOM**

When you use the $\frac{\text{part}}{\text{whole}} = \frac{\%}{100}$ rule, the important part is getting the numbers in the right positions. Certain words in the problem can signal this for you. The word "of" usually precedes the whole amount, and the word "is" can generally be found near the part. Some people remember the rule as $\frac{\text{is}}{\text{of}} = \frac{\%}{100}$.

Let's walk through some examples of the different types of problems. First, finding a percentage.

What percentage of 58 is 22?

Look for "of." 58 is the whole. Look for "is." 22 is the part.

$$\frac{\text{part}}{\text{whole}} = \frac{\%}{100}$$

$$\frac{22}{58} = \frac{x}{100}$$

$$58x = 2200$$

$$x \approx 37.9$$

22 is about 37.9 percent of 58.

You might round an answer like that to 38 percent if you don't want to deal with the decimal. It will depend on the work you're doing. If you were going to compare two percentages that were both close to 38 percent, you'd want the decimals so you could see which was bigger. If you want to know how much your savings increased, 38 percent is probably just as informative as 37.9 percent.

Let's look at another example. In this one, you'll find the whole.

46 is 27% of what number?

Look for "of." The whole is "what number," which means it is unknown. Look for "is." 46 is the part. Both 46 and 27 are near the "is" but 27 has the % sign, so you know it is the percentage.

$$\frac{\text{part}}{\text{whole}} = \frac{\%}{100}$$

$$\frac{46}{x} = \frac{27}{100}$$

$$27x = 4600$$

$$x \approx 170.4$$

46 is 27% of 170.4, approximately. Again, how much you round will depend on the situation.

One last example before we move on. This one asks you to find the part.

What is 83% of 112?

Look for "of." 112 is the whole. Look for "is." "What" is the part, that is, the part is unknown. The percent sign tells you that 83 is the percentage.

$$\frac{part}{whole} = \frac{\%}{100}$$

$$\frac{x}{112} = \frac{83}{100}$$

$$100x = 112 \times 83$$

$$100x = 9296$$

$$x = 92.96$$

83% of 112 is 92.96. Because that decimal terminates after only two places, you can give the exact answer and not bother with rounding.

If you feel like those examples all seemed very much the same, you're right. The process is exactly the same: cross-multiply and divide. The difference is the piece that's unknown, and, of course, the numbers themselves.

Percentages can be greater than 100 percent, and that often happens when you turn a question around. You can say that 2 is 50 percent of 4, or you can reverse the comparison and say 4 is 200 percent of 2. 100 percent is the whole thing, so 200 percent is the whole thing and the whole thing again, or twice as much. The good news is you do the three types of problems exactly the same way even if the percentages are greater than 100 percent.

**CHECK POINT**

11. 45 is 20 percent of what number?

12. 16 is what percentage of 64?

13. What is 15% of 80?

14. 63 is what percentage of 21?

15. What is 120 percent of 55?

# Calculating with Percentages

Because percentages are really ratios with a denominator of 100, they can easily be changed to fractions or decimals. To change a percentage like 44% to a fraction, just remember that percent means "out of 100." $44\% = \dfrac{44}{100}$, which simplifies to $\dfrac{11}{25}$. To change a percentage to a decimal, drop the percent sign and move the decimal point two places to the left. $36.8\% = 0.368$ and $3\% = .03$.

If you have a fraction and want to change it to a percentage, you can use a proportion as you did earlier, or, if you can conveniently change it to a denominator of 100, you can let it tell you what percent it is. The fraction $\dfrac{1}{4} = \dfrac{25}{100}$ so it equals 25%, but $\dfrac{6}{13}$ doesn't easily convert to a denominator of 100, and it would be better to use the proportion method.

To change a decimal to a percentage, move the decimal point two places to the right and add a percent sign. $0.15 = 15\%$ and $8.93 = 893\%$.

Yes, you can have a percentage greater than 100 percent. 100 percent is the whole thing. Percentages greater than that say you have more than the whole thing. If this week you earn 150 percent of what you earned last week, you earn all of what you earned last week (100%) and another half of that (50%).

---

**CHECK POINT**

16. Change 42% to a fraction.

17. Change 85.3% to a decimal.

18. Change $\dfrac{5}{2}$ to a percentage.

19. Change 0.049 to a percentage.

20. Change 5.002 to a percentage.

---

## Calculating Interest

Much of the time, when you encounter percentages, it's in some kind of financial calculation. It might be a big deal, like the interest you'll pay on your mortgage, or the interest the bank will pay you on your savings. Or it might just be leaving an appropriate tip for your server when you eat out.

*Interest* is money paid for the use of money. If you borrow money, you must pay it back plus additional money as a fee for the loan. If you deposit money in a bank account or invest money, you receive interest for allowing the institution to use your money. This cost of the loan is usually calculated as a percent of the amount you borrowed.

The amount borrowed or invested is the *principal*, and the percent of that principal that will be charged (or paid) in interest every year is the *rate*, sometimes called the APR, for annual percentage rate. To find the total interest on a loan or investment, you multiply the principal times the rate times the time, measured in years. The formula for simple interest is $I = Prt$, where $I$ stands for interest, $P$ for principal, $r$ for rate, and $t$ for time.

> **DEFINITION**
>
> **Interest** is money you pay for the use of money you borrow or money you receive because you've put your money into a bank account or other investment.
>
> The **principal** is the amount of money borrowed or invested. The **rate** is the percent of the principal that will be paid in interest each year.

Imagine that you open a new bank account with a deposit of $4,500, and the bank pays 1.5 percent interest per year. You leave that money in the account and don't add any money to it or take any money out for a total of 6 years. How much interest will you earn? And how much will be in your account after those 6 years?

Use the formula $I = Prt$, with $I$ as the interest, $P$ as the principal or amount you deposited, and $t$ as 6, the time in years. Take the rate of 1.5% per year and change the percent to a fraction or decimal. You will earn I = 4,500 × 0.015 × 6 = 405. You'll earn $405 interest, which will be added to your $4,500, so that after 6 years you'll have $4,905.

> **WORLDLY WISDOM**
>
> The $I = Prt$ formula is the calculation of simple interest. Many banks pay what's called compound interest. They calculate your interest every year (and sometimes more often) and add it to your account. The next year, if you don't disturb the money, you get interest on your original principal and on the interest. It goes on like that over and over, and each time interest is calculated you have a little more money in the bank, and you get a little bit extra interest. Over time, it can amount to quite a bit of extra money. The formula for compound interest is quite a bit more complicated and requires exponents.

If you want to plan for a special event, you might want to know how much you should deposit to earn enough interest to pay for your vacation three years from now. Or you might want to know how long it would take to earn a certain amount of interest. Those questions can be answered with the same simple interest formula, but a slightly different calculation.

Suppose you know that you need $2,000 to pay for a vacation three years from now, and you'd like that $2,000 to be the interest you earned on your savings in a bank that pays you 2 percent interest. Start with the formula $I = Prt$ and fill in what you know. Interest is 2,000, rate is 0.02, and time is 3 years.

$I = Prt$

$2,000 = P \times 0.02 \times 3$

Do what arithmetic you can on the right side.

$2,000 = P \times 0.06$

Then find $P$, the amount of principal you need to make that happen, by dividing 2,000 by 0.06. Don't be discouraged when you find that you need about $33,333.33.

What if you could only deposit $5,000? How long would it take to earn $2,000 in interest? The easy answer is "a lot longer" but how long? Set up the formula with what you know. You want $2,000 in interest, you want to deposit a principal of $5,000, and the bank will pay you 2 percent per year.

$I = Prt$

$2,000 = 5,000 \times 0.02 \times t$

$2,000 = 100 \times t$

$2,000 \div 100 = t$

It will take 20 years. Patience will be required.

## Calculating Tax

Most countries ask their citizens to pay income tax, a percent of what they earn, to fund the government. Most states have a sales tax, a percentage of each sale that goes to fund local government. There are a great many rules about what is and isn't subject to these taxes and under what conditions you might not have to pay them, and every country or state sets its own rate. It legislates what percent of your income or the purchase price you must pay in tax.

Whatever differences there may be, the tax is a percent of your income or a percent of your purchase. If you go out shopping and spend $273 in Colorado, the state sales tax will be 2.9 percent of your purchase. 2.9% of $273 is $0.029 \times 273 = 7.917$, which rounds to $7.92. That would be added on to your bill of $273, bringing your total to $273 + $7.92 = $280.92.

If you made the same purchase in Illinois, the state would ask for 6.25 percent, so your tax would be $0.0625 \times 273 = 17.0625$, which rounds to $17.06. That would make your bill $273 + $17.06 = $290.06.

When you go out shopping, you don't always know exactly what the sales tax is. If you buy a jacket with a price tag that says it costs $78.95, and the cashier asks you for $83.81, you can subtract 83.81 − 78.95 = 4.86 and know that you paid $4.86 in sales tax. But what is the tax rate?

Sales tax = rate × amount of the sale, so the rate will equal the sales tax divided by the amount of the sale. Take the $4.86 you paid in tax and divide by the $78.95 cost of the jacket.

4.86 ÷ 78.95 ≈ 0.0616, or 6.16%.

> **MATH TRAP**
>
> The amount you pay in tax on a sale is not always just the state sales tax. Many cities add their own sales tax in addition to the state sales tax. The city sales tax will also be a percent of your purchase, but because both are added on to your bill, you may not be able to tell how much went to the city and how much to the state.

## Calculating Tips

In many U.S. restaurants, the staff who serve the customers' meals are paid a lower than usual wage and depend on tips to make up the rest of their income. Tipping began as a way to reward excellent service but has become an expected addition to the bill in many restaurants. So the question becomes, "How much should I tip?" Usually, the recommendation is that you tip your server a percentage of the bill for the meal. The recommended percent varies by region, by the type of restaurant, and by the quality of service, but is usually in the range of 15 to 20 percent.

If you and a friend have lunch at a restaurant, and the bill for your food and drinks totals $27.46, how much of a tip should you leave for your server?

If you feel that 15 percent is appropriate, you would multiply the bill amount by 0.15. 27.46 × 0.15 takes a bit of work, but it equals $4.12. If you'd rather leave a 20 percent tip, multiply 27.46 × 0.20, and the tip will be $5.49. With that information in hand, you can decide to leave somewhere between $4 and $5.50. Most people won't worry about the pennies and will figure the tip based on the whole dollar amount.

> **WORLDLY WISDOM**
>
> People who know the sales tax rate in their area often use the tax to help them calculate the tip. If you know your sales tax is 5 percent of the bill, you can look at your check to see what the tax amount is. If you want to tip 15 percent, triple the tax. If you want to tip 20 percent, multiply the tax by 4. A tax rate of approximately 6 percent is fairly common, and many people triple the tax to get 18 percent of the bill. That's nicely in the 15 to 20 percent range.

To find 15 or 20 percent quickly by mental math, remember that 10 percent of a number is one tenth of the number, or the number divided by ten. Dividing by ten is just moving the decimal point one place left. 10 percent of $27.46 is $2.75 (rounded up from $2.746). Double that to find 20 percent, and the tip is $5.50, or if you want 15 percent, take half of $2.75, which is about $1.38, and add that to the $2.75, to get $4.13.

---

### CHECK POINT

21. Find the simple interest on $18,000 invested at a rate of 4% for 5 years.

22. If you paid $130 simple interest on a loan of $1,000 for 2 years, what was the interest rate?

23. How much tax will you pay on a purchase of $175 if the sales tax rate in your area is 4.7%?

24. If the dinner bill for your family is $35.84, and you want to leave a 20% tip for your server, what is the amount of the tip you should leave?

25. A restaurant that sells a fixed price dinner for $22 per person tells you that for groups of 8 or more, they will automatically add an 18% tip to the bill. If you and seven friends plan to go for dinner, how much should each person be prepared to pay for the dinner and the tip?

## Percent Increase and Percent Decrease

You'll sometimes see an advertisement that claims there's been a 40 percent increase in the number of vitamins in a bottle, or that there's been a 30 percent decrease in the price of something. These are examples of percent increase or percent decrease. They compare the change, whether increase or decrease, to the original amount. And it's amazing how many times those ads have the calculations wrong.

There's no reason for you to get them wrong. The original amount is the whole, and the change is the part. To calculate a percent increase or a percent decrease:

- Identify the original amount. That's the whole.

- Calculate increase or decrease. That's the part.

- Use $\dfrac{\text{part}}{\text{whole}} = \dfrac{\%}{100}$ to calculate the percentage.

Let's start with an example of a percent increase.

Allison invests $800 in a stock she researched. After a year, her investment is worth $920. What is the percent increase in the value of her investment?

Original: $800

Increase: $920 − $800 = $120

$$\frac{\text{part}}{\text{whole}} = \frac{\%}{100}$$

$$\frac{120}{800} = \frac{x}{100}$$

$$800x = 12,000$$

$$x = \frac{12,000}{800} = \frac{120}{8} = 15$$

The value of her investment increased by approximately 15%.


Here's another example. This time it's a decrease, but your method is the same.

Melissa buys $200 worth of collectibles at a flea market and tucks them away, hoping they will increase in value. Unfortunately, when she tries to sell them, she finds they are only worth $175. What is the percent decrease in the value of her investment?

Original: $200

Decrease: $200 − $175 = $25

$$\frac{\text{part}}{\text{whole}} = \frac{\%}{100}$$

$$\frac{25}{200} = \frac{x}{100}$$

$$200x = 2500$$

$$x = 12.5$$

Her investment decreased approximately 12.5%.

Ready to make sure you've got it? Here are some for you to try.

**CHECK POINT**

26. George invests $500 in stock and later sells the stock for $650. By what percent did his investment increase?

27. Find the percent decrease in Sylvia's mile time if it was 7.5 minutes when she began training and now is 6.75 minutes.

28. Paolo adopted a shelter dog who weighed 8 pounds, but by the next visit to the vet, the dog weighed 8.5 pounds. Find the percent change in the dog's weight.

29. Find the percent decrease in Shawn's weight if he was 180 pounds when he went out for the team and now is 150 pounds.

30. Your favorite brand of ice cream usually comes in a container that holds 1.5 quarts. If the company offers a special container that holds 2 quarts, what is your percent increase in ice cream?

## The Least You Need to Know

- A ratio compares numbers by multiplying or dividing. A proportion is two equal ratios. A percentage expresses a ratio by comparing to 100.

- In any proportion, the product of the means is equal to the product of the extremes.

- Solve a proportion by filling in the known quantities and a variable for the unknown, cross-multiplying, and dividing.

- To solve percentage problems, use the rule $\frac{\text{part}}{\text{whole}} = \frac{\%}{100}$.

- Simple interest is calculated by multiplying the principal times the rate times the time. Taxes and tips are calculated as a percentage of the total amount.

# PART 2

# Into the Unknown

It's nice to take a world tour, even in the world of numbers, but sometimes it can be a scary trip. If you don't know where you're going, or you think you do but find yourself lost, you can feel unsettled. Whether you're in a foreign country or in the middle of solving a problem, realizing that you're lost is a frightening feeling. The best way to prevent that situation is good preparation, so that's what we aim to do.

In this part, you'll prepare to deal with those situations in which one or more of the numbers in your problem is unknown. You'll look at ways to find those missing numbers, or if that's not possible, at least to organize the possibilities. Just as having a good map and a good plan will keep you from getting lost, having a clear picture and knowing what steps to take will help you through a problem. You're about to draw the map that will let you find your way into algebra.

# CHAPTER 8

# Variables and Expressions

One of the markers of the shift from arithmetic to algebra is the introduction of variables. In arithmetic, you know the numbers and you know what operations need to happen. In algebra, you step into the world of the unknown. You know some, but not all, of the numbers, and you know the result of the operations. A lot of your work is going backward to try to fill in the unknown pieces. Until you figure out what they are, those unknown numbers are represented by variables.

In this chapter, you'll learn what variables are and how to write phrases and sentences that use them. When you start using variables, it's important to consider what kind of number the variable might be standing in for. Those numbers are called the domain of the variable. This is also a time to look at some of the arithmetic you can do with variables and learn where you need to be especially careful and why.

## In This Chapter

- Writing expressions with variables
- Finding values for variables
- Multiplying with the help of exponents
- Dividing with variables

# Using Variables

A *variable* is any symbol that stands for a number. In algebra, the symbol is usually a letter, but that doesn't have to be the case. You might have looked at questions that said $3 + ? = 5$ and you understood that you were supposed to find the number that replaced the $?$ to make the statement true. In that case, $?$ is a variable. You might have seen puzzles that said $\triangle + \diamond = 9$. There are many different right answers for that one, but the $\triangle$ and the $\diamond$ are acting as variables.

> **DEFINITION**
>
> A **variable** is a letter or symbol that takes the place of a number.

A variable is used to take the place of a number because the value of the number is unknown, or because the number that goes in that place may be changing. Perhaps a pattern is being represented in which different values are possible.

For example, imagine you are taking a test with 20 questions. Each question is worth 5 points, so if you get all the questions correct, you earn 100 points. But what if you don't get them all right? Your teacher might use a rule or pattern that says your score is 5 points times the number of questions you answer correctly. Using the variables $S$ for your score and $n$ for the number of questions you answer correctly, you can write the rule as $S = 5{\times}n$. That rule will apply to everyone who takes the test, but each person may get a different number of questions correct. The variable $n$ can have many different values. It varies. That's where the name variable comes from.

Suppose on a different test you got your paper back with a grade of 87 on it. You want to know how many points each question was worth. You can find out by counting the number of questions you answered correctly and using the variable $p$ to stand for the point value of one question. If you answered 29 questions correctly, you can say $29{\times}p = 87$. Here you're using a variable because you don't know the number of points. The variable stands for a number that is unknown.

## The Language of Variables

When you start to use variables in your phrases and sentences, it's a lot like learning a new language. You need to learn your vocabulary, understand the grammar of the language, including the idioms, and practice, practice, practice.

Let's start with parts of speech: nouns, verbs, and such. In algebra, numbers and variables are your nouns, and symbols like $=$, $<$, and $\geq$ serve as verbs. Operation signs, like $+$ and $-$, act as conjunctions, the way "and" and "or" would in English.

To translate algebra into English, you can usually just read the symbols and make a few shifts, like saying "and" for $+$ and "is" for $=$. The sentence $x + 3 = 8$ is the algebra language equivalent of "some number and three combine to make eight." The $x$, the 3, and the 8 are nouns. The $x$ and the 3 are connected by the conjunction +, and the verb, the action of the sentence, is the =. The = says "is," or "makes."

The difference between a phrase and a sentence is whether or not you have a verb. Sentences have verbs, phrases do not. $x + 9$ is a phrase. It has no equal sign or inequality sign. It sounds like "a number increased by 9." On the other hand, $x + 9 > 21$ is a sentence. The $>$ is the verb "is greater than." It says "some number increased by 9 is greater than 21."

When parentheses appear, it's often easier to read if you refer to "the quantity." The sentence $5(x + 7) = 45$ can be read as "five times the quantity $x$ plus 7 is 45." The words "the quantity" tell your listener that parentheses are grouping some symbols into one quantity.

The larger your vocabulary, the better able you are to say what you really mean, so let's look at some common language. That sentence $5(x + 7) = 45$ could also be read as "five times the sum of $x$ and 7 is 45." A sum is the result of an addition. If you change the plus sign to a minus sign, $5(x - 7) = 45$ can be read as "five times the difference of $x$ and 7 is 45." A difference is the result of a subtraction. Whenever you talk about a difference, you read the numbers in the order they appear. The difference of $x$ and 7 is $x - 7$. The difference of 7 and $x$ is $7 - x$.

## Idioms in Algebra

Idioms are phrases used in a language that just can't translate word for word. If I tell you to give me break, I'm not asking you to hurt me. I'm telling you to stop teasing me or trying to fool me. "Give me a break" is an idiom.

Algebra has idioms, too. One in particular has to do with subtraction. The phrase "6 less than some number" sounds as though it would begin with a 6 if you wrote it in symbols, but it actually translates to $x - 6$. To see why, put a number in place of the words "some number." If I say "6 less than 25," you think of $25 - 6$, or 19. You take 6 away from the number. So 9 less than some number is $t - 9$, and 83 less than some number is $y - 83$.

 **MATH TRAP**

Don't confuse "6 less than some number" with "6 is less than some number." The word is signals a verb. "6 less than some number" translates to $x - 6$, but "6 is less than some number" would be $6 < x$.

If you want to read the phrase $7(x + 5)$, you could say "seven times the quantity $x$ and 5," or you could say "the product of 7 and the quantity $x$ plus 5." A product is the result of multiplication. When you want to say something about division, like $\dfrac{x}{5} = 11$, you can say "a number divided by 5 is 11," or you can say "the quotient of $x$ and 5 is 11." A quotient is the result of division. The order in which the numbers are named tells you which is the dividend and which is the divisor. The quotient of $x$ and 5 is $x \div 5$, but the quotient of 5 and $x$ is $5 \div x$.

**CHECK POINT**

Write each sentence using variables.

1.  The quotient of some number $x$ and twelve is greater than four.

2.  The difference of a number $t$ and nineteen is twenty-two.

3.  The sum of a number $n$ and the quantity $n$ increased by three is one hundred fifty-four.

4.  The product of a number $y$ and the quantity five less than $y$ is eighty-four.

5.  The quotient of a number $p$ and the quantity $p$ increased by one is two.

# Multiplying with Variables

If variables take the place of numbers, then it makes sense that you should be able to do arithmetic with variables, because you do arithmetic with numbers. Not knowing what number the variable stands for is an obvious problem when it comes to arithmetic, but there are some arithmetic operations you can do with variables, if you work carefully.

Let's start with multiplication, because that's probably the most common operation and the one with the fewest restrictions or dangers. If you need to multiply a constant, like 4, by a variable, like $t$, you can do that. Because you don't know what number $t$ stands for, you can't give a number as the answer, but you can write $4t$.

If you needed to multiply a constant, like -7, by the product of a constant and a variable, like $4t$, you apply the associative property. $-7 \cdot (4t) = (-7 \cdot 4) \cdot t = -28t$. You multiply the constant by the coefficient and keep the variable as the final factor.

Things get interesting when a multiplication involves more than one variable. If you wanted to multiply $-7t$ by $4t$, you could call on the associative property and the commutative property. $-7t \cdot 4t = (-7 \cdot t) \cdot (4 \cdot t) = (-7 \cdot 4) \cdot (t \cdot t) = -28 \cdot t \cdot t$. Do you remember the shortcut for writing repeated multiplication, like $t \cdot t$? You can use an exponent. $-28 \cdot t \cdot t = -28t^2$.

## Working with Exponents

Exponents, you remember, are symbols for repeated multiplication. The expression $5^3$, for example, means $5 \cdot 5 \cdot 5$. That ability to express repeated multiplication with exponents is even more important when you're working with variables, because you don't know what number the variable stands for, so you can't evaluate the multiplication. When you write $b^n$ you say that you want to use $b$ as a factor $n$ times. The expression $b^5$ means $b \cdot b \cdot b \cdot b \cdot b$, but until you know what number $b$ stands for, that's all you can say.

When you work with exponents, there are three basic rules to remember.

**Multiplication:** When you multiply powers of the same base, keep the base and add the exponents.

This means that $x^2 \cdot x^3 = x^5$ because $x^2 \cdot x^3 = (x \cdot x) \cdot (x \cdot x \cdot x) = x^5$. Don't forget that when you have a single variable, like $y$, even though you don't see an exponent, $y$ is $y^1$. Multiplying $y \cdot y^3$ gives you $y^4$. Multiplying a power of a variable by the variable again raises the exponent.

**Division:** When you divide powers of the same base, keep the base and subtract the exponents.

This means that $\dfrac{x^8}{x^6} = x^2$. Don't forget that a variable to the zero power equals 1.

**Powers:** When you raise a power to a power, keep the base and multiply the exponents.

This one doesn't come up as often, but it means that $(t^7)^3 = t^{21}$. Remember that exponents work only on what they touch, so $3x^2$ means $3 \cdot x \cdot x$, because the exponent is only touching the $x$, but $(3x)^2 = (3x) \cdot (3x)$ because the exponent is touching the parenthesis, and so the whole quantity is squared.

## Ways to Multiply

You can multiply a constant by a variable or a power of a variable: $-5 \cdot x^2 = -5x^2$.

You can multiply a constant by a product of a constant and a variable: $13 \cdot 4t = 52t$.

You can multiply a variable or a power of a variable by a variable or a power of a variable: $y \cdot y^3 = y^4$.

You can multiply a variable or a power of a variable by a product of a constant and a variable or a power of a variable.: $a^2 \cdot 6a^3 = 6a^5$.

You can multiply the product of a constant and a variable (or a power of a variable) by a product of a constant and a variable (or a power of a variable): $-8x^2 \cdot 4x^2 = -32x^4$.

And you can even multiply a constant, a variable, or the product of a constant and a variable by an expression: $2x(3x - 5) = 2x(3x) - 2x(5) = 6x^2 - 10x$ and $3t^2(t^2 + 2t + 1) = 3t^2(t^2) + 3t^2(2t) + 3t^2(1) = 3t^4 + 6t^3 + 3t^2$.

In case you were wondering, you can multiply with two different variables. You just simplify as much as possible: $5x(2y) = 10xy$ and $2ab^2(3a^2b) = 2 \cdot 3 \cdot a \cdot a^2 \cdot b^2 \cdot b = 6a^3b^3$.

---

**CHECK POINT**

Put each product in simplest form.

6. $5(-6a)$

7. $x(4x^2)$

8. $2y(-5y^3)$

9. $3t^2(-8t^5)$

10. $7z(3z + 5)$

# Dividing with Variables

When you try to divide by a variable or an expression that contains a variable, you have the same problem as you did for multiplication: you don't know what number the variable stands for. Just as it did in multiplication, that means you can't always do as much simplifying as you might like to, but in division, it also causes another problem. You can multiply by any number, but you can't divide by zero. If you don't know what number the variable stands for, you don't know what you're dividing by, and because division by zero is undefined, that's dangerous.

That's why whenever you divide by a variable or write a fraction with a variable in the denominator, you always include a note that says, in one way or another, "as long as this isn't zero." If you wanted to divide $5x^2 \div x$, you'd include a little warning: $x \neq 0$.

## Consider the Domain

When we talk about variables in algebra, we often talk about the domain of the variable. The domain is the set of all values that you can reasonably substitute for the variable. It's all the values for the variable that make sense.

One of the first things you look at when you think about the domain is dividing. If you're trying to divide by the variable, 0 can't be in the domain. There are other problems to be aware of as well. You can't take the square root of a negative number, for example, and if you're using your variable to solve a word problem, only numbers that make sense in the problem should be in the domain. The length of a fence can't be -10 feet, and you can't invite 12.4 people to your party. But the number one concern is making sure you never divide by zero.

## Rules for Dividing with Variables

So you have to be careful and include a warning note, but how do you divide by a variable or an expression containing a variable? There are several ways, because as the expressions get more complicated you need different methods, but the basic rules come back to fractions and exponents.

If you have a product of constants and variables divided by a product of constants and variables, you can think of the problem as a fraction that needs to be put in simplest form. Let's look at some examples.

If you want to divide $8x$ by 2, you can think of it as $\frac{8x}{2} = \frac{8}{2} \cdot \frac{x}{1} = \frac{4}{1} \cdot \frac{x}{1} = 4x$. You're just dividing the coefficient of the numerator by the constant in the denominator. Now it's not hard to think of a case, like $\frac{8x}{3}$, where it might be better to just leave it as it is, or $\frac{y}{5}$ where there's nothing you can do. But if the denominator is a constant, and you can cancel that constant with a coefficient in the numerator, that's what you want to do.

What if there's a variable in the denominator? To divide $\dfrac{6t^2}{t}$, add your warning note, $t \neq 0$, and then let the coefficient be and focus on the variables. This is where the rules for exponents come in. $\dfrac{6t^2}{t} = \dfrac{6}{1} \cdot \dfrac{t^2}{t} = \dfrac{6}{1} \cdot \dfrac{t}{1} = 6t$, provided $t \neq 0$. Unless the variable in the denominator is different from the variable in the numerator, you should be able to do some simplifying with problems like this.

The next step is putting the pieces together. To divide $\dfrac{27\,y^5}{54\,y^3}$, start with the fact that $y \neq 0$ and then group coefficients with coefficients and variables with variables. $\dfrac{27\,y^5}{54\,y^3} = \dfrac{27}{54} \cdot \dfrac{y^5}{y^3}$. Reduce the fraction, and use the rules for exponents to simplify the variable part. $\dfrac{27\,y^5}{54\,y^3} = \dfrac{27}{54} \cdot \dfrac{y^5}{y^3} = \dfrac{1}{2} \cdot \dfrac{y^2}{1}$. You can write your final answer as $\dfrac{y^2}{2}$ or $\dfrac{1}{2}\,y^2$, whichever you prefer, but remember to include $y \neq 0$.

## Dividing with Other Operations

When addition or subtraction slips into the problem, things get a little more complicated. To divide $\dfrac{25x^5 + 70x^4 + 15x^3}{5x^3}$, you need to break it into three fractions. Each piece of the numerator is going to be divided by the denominator.

Think for a minute about adding fractions, just regular fractions, like $\dfrac{1}{2} + \dfrac{2}{3}$. First you have to change each fraction to an equivalent fraction with a common denominator, in this case, 6. $\dfrac{1}{2} + \dfrac{2}{3} = \dfrac{3}{6} + \dfrac{4}{6}$. Then you'll add the $3 + 4$ and put the answer over the common denominator of 6.

What you're going to do with this variable division problem is step to when you had separate fractions with the same denominator. $\dfrac{25x^5 + 70x^4 + 15x^3}{5x^3}$ is going back to $\dfrac{25x^5}{5x^3} + \dfrac{70x^4}{5x^3} + \dfrac{15x^3}{5x^3}$. Here's how it works. Don't forget: $x \neq 0$.

$$\frac{25x^5 + 70x^4 + 15x^3}{5x^3} = \frac{25x^5}{5x^3} + \frac{70x^4}{5x^3} + \frac{15x^3}{5x^3}$$

Simplify each of the three fractions by grouping coefficients with coefficients, reducing the fraction, and then grouping variables with variables, using the rules for exponents.

$$\frac{25x^5 + 70x^4 + 15x^3}{5x^3} = \frac{25x^5}{5x^3} + \frac{70x^4}{5x^3} + \frac{15x^3}{5x^3}$$
$$= \frac{25}{5} \cdot \frac{x^5}{x^3} + \frac{70}{5} \cdot \frac{x^4}{x^3} + \frac{15}{5} \cdot \frac{x^3}{x^3}$$
$$= \frac{5}{1} \cdot \frac{x^2}{1} + \frac{14}{1} \cdot \frac{x}{1} + \frac{3}{1} \cdot \frac{1}{1}$$
$$= 5x^2 + 14x + 3$$

So $\dfrac{25x^5 + 70x^4 + 15x^3}{5x^3} = 5x^2 + 14x + 3$, as long as $x$ is not zero.

The toughest problems are the ones in which there is addition or subtraction in the denominator. Unfortunately, they're not as easy to break into smaller problems. If you go back to regular fractions for a minute, remember that $\dfrac{5}{2+3}$ does not equal $\dfrac{5}{2}+\dfrac{5}{3}$.

There are ways to do division problems like $\dfrac{x^2-5x+6}{x-3}$, but you'll need more experience with different algebra tools before you're ready for them. For now, practice with division problems in which the denominator does not include addition or subtraction. Reduce fractions, use the laws of exponents, and don't forget to always add the note that says your denominator can't be zero.

**CHECK POINT**

Write each quotient in simplest form.

11. $\dfrac{-14t}{7}$

12. $\dfrac{11a^3}{a^2}, a \neq 0$

13. $\dfrac{18x^4}{6x^3}, x \neq 0$

14. $\dfrac{42y^7}{14y^4}, y \neq 0$

15. $\dfrac{12x^3-72x^2+32x}{4x}, x \neq 0$

## The Least You Need to Know

- Variables are letters or symbols that represent unknown numbers.
- Multiply numbers with numbers and variables with variables.
- When multiplying the same variable many times, use an exponent.
- To divide an expression by a variable, make separate fractions and reduce.
- You cannot divide another number by zero.

# Adding and Subtracting with Variables

The shift from arithmetic to algebra begins with the introduction of variables. Almost immediately, you have to start doing arithmetic with variables. The rules of arithmetic don't change, but working with one or more unknown quantities requires some new strategies. If you have four 3s, you can do the multiplication and know you have 12, but if you have four $x$'s, all you can do is say that you have $4x$. What $4x$ is worth depends on what number $x$ stands for.

The statement that you have four $x$'s, written as $4x$, is a simple form of multiplication. As you saw in the last chapter, more complicated multiplication requires multiplying the coefficients, the numbers that tell you how many you have, and multiplying the variables, which usually requires using exponents.

In this chapter, we'll look at more arithmetic with variables, with a focus on adding and subtracting. I'll explain when you can, and when you can't, add or subtract expressions involving variables, and how to do it when you can. And I'll introduce you to polynomials, a family of expressions built by adding variable terms.

## In This Chapter:

- Understanding terms
- How to identify like terms
- Adding and subtracting like terms
- Recognizing when an expression is in simplest form
- How to name polynomials

# When Are Terms "Like Terms"?

You're often going to see the word *term* in this chapter, so let's start by making sure you understand what it means. A *term* is a constant, a variable, or an expression involving multiplication of constants and variables. You use the more general word *expression* to cover just about anything you write using numbers and variables. Terms are a particular subset of expressions that involve only multiplication.

📖 **DEFINITION**

A **term** is an algebraic expression made up of numbers, variables, or both that are connected only by multiplication.

Any number on its own is a term. Constants, like 1 or -7, are terms. Any variable, such as $x$ or $y$, is a term. When you multiply numbers and variables, you get terms like $-4y$, or $xy$, or $x^2$, or $18xy^2$. You can have numbers, variables, numbers and variables multiplied together, and variables multiplied together, which may give you exponents. That's all okay in a term. You just can't add or subtract, divide by a variable, or have a variable under a square root sign. Dividing by a constant is the same as multiplying by a fraction, so that's allowed.

## Simplify to Find Terms

It's possible that an expression may not look at first like it fits the definition of a term, but you might find that you can simplify the expression and the simplified form does fit the definition of term.

For example, the expression $5\sqrt{x^2} + \dfrac{6x^2}{x}$ seems to break all the rules. There's addition, there's a variable under a square root sign and there's division by a variable. But you can do that division. $\dfrac{6x^2}{x} = 6x$ so you can make the expression $5\sqrt{x^2} + \dfrac{6x^2}{x} = 5\sqrt{x^2} + 6x$.

Then you might notice that $\sqrt{x^2} = x$, so $5\sqrt{x^2} + \dfrac{6x^2}{x} = 5\sqrt{x^2} + 6x = 5x + 6x$.

$5x + 6x$ is $11x$, and that's a term. Take a minute to think about whether you can simplify an expression before you decide if it fits the definition.

✏️ **CHECK POINT**

Decide if each expression is a term.

1. $4x$ ✓

2. $-12$ ·

3. $-2t^7$.

4. $\dfrac{6}{y}$

5. $\dfrac{a}{6}$

# When Can You Combine Terms?

Terms may only involve multiplication, but you need to think about what happens when you want to do something besides multiply, specifically, when you want to add or subtract terms. When the terms are just numbers, addition and subtraction are straightforward: $4 + 8 = 12$ and $-5 + 9 = 4$. You're just adding (or subtracting) according to the rules of arithmetic. The moment variables enter the picture, however, you're faced with a dilemma. How do you add two numbers if you don't know what they are?

You may already have part of the answer. If you add $x + x$, whatever number $x$ stands for, you have two of that, so $2x$. But what if you need to add $x + y$? You don't have two $x$'s and you don't have 2 $y$'s. You can't really say much about what you do have, except that you have $x + y$.

Different variables, like $x$ and $y$, are *unlike terms*. They're different. It's an apple-and-orange kind of thing. One apple plus one orange doesn't give you two apples or two oranges or two appleoranges. It can give you two fruits, if you take a common denominator approach to the matter, but when you're working with variables, you don't know enough about them to find a common denominator. You're stuck admitting that unlike terms can't be combined.

**DEFINITION**

**Unlike terms** are terms with different variables, such as $x$ and $y$.

In order to combine terms, they must have the same variable. They must be like terms. As usual, there are some complications to that simple rule. What about $x + xy$? Or $x + x^2$? In each of these examples, the second term has an $x$, but it also has something else. Can you combine $x$ with $xy$ (which is $x$ multiplied by $y$) or $x^2$ (which is $x$ multiplied by $x$)?

If you could combine $x$ and $xy$, what would it give you? The $xy$ isn't the same number as $x$ (unless $y$ happens to be 1, and you can't bet on that) so you don't have $2x$, and you can't just throw the $y$ away. The terms $x$ and $xy$ are unlike terms.

The $x^2$ may look more like the $x$, but it's not the same. If $x$ were equal to 5, $x^2$ would be 25, and they'd add to 30. That's six times the value of $x$. But if $x$ were equal to 3, $x^2$ would be 9, and they'd add to 12, which is four times the value of $x$. There are so many possibilities, and while there is a pattern to them, it's not immediately clear. It's difficult to say what $x + x^2$ equals because $x$ and $x^2$ are not like terms.

What is the pattern to $x + x^2$?

If $x = 1$, $x + x^2 = 2$ or $2x$.

If $x = 2$, $x + x^2 = 6$ or $3x$.

If $x = 3$, $x + x^2 = 12$ or $4x$.

Can you see it? The answer is always a multiple of $x$, and the multiplier is 1 more than the value of $x$. You can write it like this: $x + x^2 = (x + 1)x$. This is called the factored form of $x + x^2$. If you use the distributive property to multiply $x(x + 1)$, you'll see that it equals $x + x^2$.

Think of it this way: $x$ stands for a number, which we could imagine as the length of a line segment. If $x$ is the length of a line segment, then $x^2$ would be the area of a square whose sides are $x$ units long. Visually, a line segment and a square are very different things. Even though both $x$ and $x^2$ have an $x$ in their names, they're very different things, different numbers. They're unlike terms.

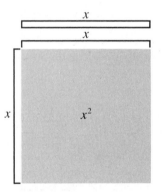

Even though they both contain the same variable, $x$ and $x^2$ represent different ideas. There's something going on in $x^2$ that's not happening in the plain $x$. Visually, you can think of the $x^2$ expanding into a square, while the $x$ is still a line segment. Numerically, the $x^2$ has multiplication going on that's not happening in the simple $x$. The $x$ and the $x^2$ are clearly related, but they're not the same. They're not like terms.

Different variables are clearly not alike, but even terms built from the same variable may be different from one another. To be *like terms*, terms must have the exact same variable and the exact same exponent. Their variable parts are exact matches. Only their coefficients are different. That means $4y^3$ and $-7y^3$ are like, because they have the same variable and the same exponent. The fact that the 4 and the -7 are different coefficients is okay. The coefficients are just telling you how many you have.

**Like terms** are terms that have the same variable, raised to the same power. For example, the terms $4x$ and $7x$ are like terms.

**CHECK POINT**

Label each pair of terms *like* or *unlike*.

6. $7y^2$ and $11y$

7. $3t^2$ and $5t^2$

8. $2x$ and $7x$

9. $-9a^2$ and $-15a^3$

10. $132x^3$ and $-83x^3$

# Adding and Subtracting Like Terms

So $x + x$ is $2x$, but $x + y$ must stay as $x + y$ because we have no way to combine the unlike terms. What about $2x + 3x$? The variable is the same, but the coefficients, the numbers in front, are different. Is that a problem? Like terms are terms that have the same variable and the same exponent. These terms meet that rule, so why is it okay for the coefficients to be different?

Think about what the coefficients tell you. You have 2 $x$'s and another 3 $x$'s. That's all $x$'s, so you can combine them. The $2x$ means $x + x$ and the $3x$ means $x + x + x$. You can put them all together and end up with a total of 5 $x$'s. $2x + 3x = 5x$.

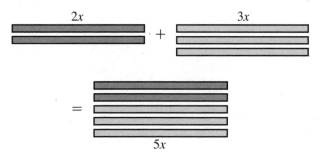

If the variable parts of two terms are identical, you can add them by just adding the coefficients. As long as you have the same variable and same exponent, you can just look at the coefficients to tell you how many you have.

There's no change to the variable part. You're just changing the count of how many of that variable you have. The variable portions tell you that you're working with the same kinds of things, and the coefficients tell you how many of them you have. If you were asked to add 7 cars and 5 cars, you'd get 12 cars, not 12 cars squared. 8 apples minus 3 apples gives you 5 apples. The cars or apples don't change. When you add or subtract like terms, only the number changes.

**WORLDLY WISDOM**

When you add fractions, you need to have like denominators. When you add terms, you need to have like terms. The denominator of a fraction tells you what kind of fraction you have, and the numerator tells you how many of them you have. If you have the same denominators, you can just add the numerators, but if the denominators are different, you can't combine the fractions. The variable part of a term tells you what kind of thing you have, and the coefficient tells you how many. If the variable parts are the same, you can add the coefficients.

If you have to subtract terms, you follow the same rule: you can only subtract like terms, and you subtract the coefficients and keep the variable part exactly as it was. To subtract $8t - 11t$, you subtract $8 - 11$ to get -3, and you keep the $t$. So $8t - 11t = -3t$.

If you're faced with an addition or subtraction problem and you realize that the terms are unlike, you just leave the problem as it is, or you might say "this cannot be combined."

**CHECK POINT**

Is it possible to complete these additions and subtractions? Complete them if you can!

11.  $-4x + 9x$

12.  $3a^2 - 2a^3$

13.  $5xy + 6xy$

14.  $120xy^2 - 80xy^2$

15.  $15z + 25x$

# Simplifying Expressions

Once you have terms, it's natural to want to start doing things with them. You can add or subtract terms if they're like terms, and you can multiply terms by other terms, like or not. You can even do both of these things, adding and then multiplying. When you start doing so many things with terms, you soon need to stop and think about how you simplify the problems you've created.

Do you need a whole new set of rules? The good news is no. Your old friend PEMDAS (Parentheses, Exponents, Multiply, Divide, Add, Subtract) will cover most situations. Let's look at a problem that's not too complicated. Suppose you wanted to do $5x(3x + 7x)$. PEMDAS says do what's in the parentheses first, so look at the $3x + 7x$ part first. Those are like terms, so you can add them and get $10x$. Your expression $5x(3x + 7x)$ becomes $5x(10x)$. Exponents are next, but there are none, so move on to multiplying and dividing. $5x(10x) = 5 \times 10 \times x \times x = 50x^2$.

How about this one? Simplify $-4xy(6x + 7y)$ using the same rules. Parentheses first. But this time, you can't add what's in the parentheses because they're unlike terms. Don't panic. You can only do what you can do, so just move on. There are no exponents, so it's time to multiply, and for that, you'll need the distributive property.

$$-4xy(6x + 7y) = -4xy(6x) + (-4xy)(7y)$$

Now, as you look at each multiplication, multiply coefficients and combine what you can. Don't be afraid to rearrange. Remember multiplication is commutative and associative.

$$-4xy(6x) = -4 \times 6 \times x \times x \times y = -24x^2y$$

$$(-4xy)(7y) = -4 \times 7 \times x \times y \times y = -28xy^2$$

Put the pieces back together to get

$$-4xy(6x + 7y) = -4xy(6x) + (-4xy)(7y) = -24x^2y + -28xy^2$$

Because you have the plus from the addition problem followed immediately by the minus sign from the -28, you can write your answer as $-24x^2y - 28xy^2$.

Your basic rules are:

- Combine what's in the parentheses if you can.

- Distribute multiplication over the addition or subtraction if you can't combine the unlike terms.

- Simplify each multiplication.

- Check the signs.

Whenever you have negatives in problems like these, it's important to be careful about the rules for signs. If the multiplier that you're distributing is positive, the signs aren't usually a problem. Terms in the parentheses that were positive will produce positive terms in the answer, and terms that were negative in the parentheses will produce negative terms in the answer. Here are some examples.

$$8x(-3a + 2b) = 8 \times (-3) \times x \times a + 8 \times 2 \times x \times b = -24ax + 16bx$$

$$5t(7t^2 - 3t) = 5 \times 7 \times t \times t^2 - 5 \times 3 \times t \times t = 35t^3 - 15t^2$$

When the multiplier is negative, however, you need to work carefully and remember that if you distribute a negative term, the signs of each term in the parentheses will change. Let's take those two examples and make the multipliers negative to see what happens.

Take the first example and change $8x$ to $-8x$. Distribute the $-8x$.

$$-8x(-3a + 2b) = -8 \times (-3) \times x \times a + (-8) \times 2 \times x \times b$$

Now that first term has a negative multiplied by a negative, so the result is going to be positive. The second term has one negative, so that result will be negative.

$-8x(-3a + 2b) = -8 \times (-3) \times x \times a + (-8) \times 2 \times x \times b = +24ax + -16bx$

You can eliminate the plus signs, condensing to $24ax - 16x$.

If you change $5t$ to $-5t$ in the second example and distribute, you get $-5t(7t^2 - 3t) = -5 \times 7 \times t \times t^2 - (-5) \times 3 \times t \times t$.

The first term will come out negative, but the second term is where you have to be careful. The product $(-5) \times 3 \times t \times t$ will give you a negative result, $-15t^2$, but you also have a minus in front of that, the one that was connecting the original terms. You get a product of $-35t^3 - -15t^2$, but the double minus becomes a plus so you end up with $-35t^3 + 15t^2$. The minus on the multiplier switches the signs.

Sometimes you'll find an expression with more than one set of parentheses and lots of terms, some like and some unlike, and some that start out unlike and then turn into like terms. You just need to take things step by step and pay attention to what is happening. Here's an example.

$-3x(6x^2 - 5) + 8x^2(4x - 9)$

There are no like terms in the first parentheses, so you can't do anything there, and no like terms in the second set of parentheses, either. Move on to multiplying. Use the distributive property.

$-3x(6x^2 - 5) + 8x^2(4x - 9)$

$= -3 \times 6 \times x \times x^2 - (-3) \times 5 \times x + 8x^2(4x - 9)$

$= -3 \times 6 \times x \times x^2 - (-3) \times 5 \times x + 8 \times 4 \times x^2 \times x - 8 \times 9 \times x^2$

$= -18x^3 + 15x + 32x^3 - 72x^2$

Now notice you have like terms: $-18x^3$ and $32x^3$. You can combine those to get a final answer of $14x^3 + 15x - 72x^2$. It's traditional to put your terms in order from highest exponent to lowest, so rewrite it as $14x^3 - 72x^2 + 15x$.

---

**CHECK POINT**

Simplify each expression.

16.  $6x(2x + 9)$

17.  $12 + 5(x + 1)$

18.  $6t^2(t - 3) - 2t^2$

19.  $5y(6y + 2) + 7y^2(4 - 12y)$

20.  $8a(2b - 5) - 2b(a - 2)$

# Polynomials

There's a particular group of expressions, which you'll hear about in algebra, that are called polynomials. The prefix *poly* means "many," so it would seem that they would be expressions with many terms, and some of them are. But the name *polynomial* is applied to any expression that fits a particular pattern, even those with only one term.

The pattern is easier to show than to describe, but let's try. Polynomials are expressions that are made by adding terms that are the product of a numerical coefficient and a power of a certain variable. For example, $8x^5$, $-6x^3$, $x^2$, and $2x$ are all terms that are the product of a numerical coefficient and a power of the variable $x$. You don't see the numerical coefficient in $x^2$ because it's 1, and we don't usually show that. Constants can also be part of a polynomial because we can say a constant, like 3, is $3x^0$.

Each of these terms could be called a polynomial all by itself. It would be a one-term polynomial, also called a monomial. *Mono* is the prefix that means "one." You can add monomials with the same variable to make a more complex polynomial. For example, $8x^5 + -6x^3 + x^2 + 2x + 3$ is a polynomial. If you add two monomials, like $5y^3 + 2y$, you make a binomial. If you add three monomials, like $t^2 + 3t + 1$, that's a trinomial. Monomials, binomials, and trinomials are all types of polynomials.

The degree of a monomial is the exponent on the variable. $8x^5$ is fifth degree, $x^2$ is second degree, and $2x$ is first degree. You don't see an exponent of 1, but that's what $2x$ really means: $2x^1$. Constants are degree zero, because we're thinking of them as a constant times $x^0$.

**MATH TRAP**

Be careful to look at all the terms of a polynomial before you decide on its degree. Don't just jump at the term that happens to be written first.

The degree of a polynomial is the highest degree of all its monomials. The polynomial $8x^5 + -6x^3 + x^2 + 2x + 3$ is a fifth degree polynomial, and $t^2 - 7t + 4$ is a second degree polynomial. For the polynomial $5t^3 - 7t + 8t^4 - 2t^2 + 5$, you need to be careful to look at the whole polynomial. The degree of the polynomial is 4, not 3. The highest degree term is in the middle, not the beginning, of the polynomial.

If there is more than one variable in a term, the degree of the term is the sum of the degrees of each variable. The term $xy$ is degree 2, one for $x$ and one for $y$. The term $3x^4y^2$ is degree 6.

**CHECK POINT**

Give the degree of each polynomial.

21. $-3a^3 + 5a^2 - 3a + 12$          24. $11y - 7y^4 + 5y^2 - 3$

22. $6 + 3b - 4b^2$          25. $6 - 4x^2 + 3x$

23. $2t - 9 + 7t^2 + 4t^3$

When you write a polynomial, no matter how many terms it has, it's traditional to write it in standard form. Standard form means that you write the terms of the polynomial in order from the highest degree monomial to the lowest. The polynomial $8x^5 + -6x^3 + x^2 + 2x + 3$ is in standard form because the monomials start with the fifth degree, then the third degree, then the second, first, and finally the zero degree term.

**WORLDLY WISDOM**

The standard form of a polynomial writes the terms in order from highest degree to the lowest.

The polynomial $4t^3 - 6t^9 + 11t - 5t^2 + 8t^7$ is not in standard form. To put it in standard form, you need to first locate the highest degree term. In this case, that's $-6t^9$. That has to be first, so move it around to the front. The $4t^3$ that was in front is positive, so you can put a plus between the ninth degree term and the $4t^3$. $-6t^9 + 4t^3 + 11t - 5t^2 + 8t^7$.

There's no eighth degree term, but there's a seventh degree, so that should be next. $-6t^9 + 8t^7 + 4t^3 + 11t - 5t^2$.

There are no sixth, fifth, or fourth degree terms, and the third degree term is already next in line. You just need to swap the last two terms. $-6t^9 + 8t^7 + 4t^3 - 5t^2 + 11t$ is in standard form.

**CHECK POINT**

Put each polynomial in standard form.

26. $a^3 + 10a^4 - 11a + 9$          29. $4 - 7m^2 + 14m^4 + 2m^3$

27. $2b^3 - 9b + 12b^2 - 5$          30. $5p - 3 + 6p^2 - 15p^3$

28. $3k^4 + 8k^5 - 13k - 7$

## The Least You Need to Know

- A *term* is a constant, a variable, or the product of a constant and a variable.
- Terms are "like" if they contain exactly the same variables, raised to the same power, such as $2y$ and $10y$.
- Only like terms can be combined.
- To add or subtract like terms, add or subtract the coefficients and keep the variable part the same.
- A polynomial is a sum of terms using different powers of the same variable. A monomial has one term, a binomial has two terms, and a trinomial has three terms.

# CHAPTER 10

# Solving Equations and Inequalities

The moment you begin to use a variable to represent an unknown number, a question arises: what is the unknown number? It's natural to want to know what number the variable represents, and in many cases, you can find out. You need to have a full mathematical sentence, called an equation or an inequality, using one variable. If you do, you can determine the value of the variable that makes the sentence true. This process is called solving the equation (or inequality).

In this chapter, you'll look at the essential steps in working back to what a variable must equal. You'll focus on how to clean up the problem before you start so you can use those fundamental techniques. Of course, there are special cases that need to be handled differently, and you'll learn how to handle those. Finally, you'll investigate what's the same and what's different if the mathematical sentence has a verb that talks about larger or smaller instead of just equal.

## In This Chapter

- Solving one- and two-step equations.
- Simplifying equations to make solving possible.
- Using a number line to solve inequalities.

# Using Equations to Find the Missing Number

Variables can stand for numbers that are unknown or numbers that change. When you use a variable to take the place of a number that changes values, you usually have an *expression*, the mathematical equivalent of a phrase.

For example, if you buy hamburgers for $3.50 each, the amount you have to pay will vary depending on how many burgers you buy. If you use the variable $h$ to stand for the number of hamburgers you buy, the amount you have to pay would be represented by the expression $3.50 \times h$. If you buy 2 burgers, $h$ is 2 and you pay $7. If you buy 10 burgers, $h$ is 10 and you pay $35. There's no one "right" value of $h$. It's a different number each time you do the problem.

On the other hand, if you know that hamburgers are $3.50 each, and you know that you spent $24.50, you can write the *equation* $3.50 \times h = \$24.50$. Now you have a full mathematical sentence, an equation, and a question: how many hamburgers did you buy? There's only one value that can take the place of $h$ and make that sentence true. Finding that value is the process of solving an equation.

> **DEFINITION**
>
> An **expression** is a mathematical phrase. Expressions may include variables, but they do not have an equals sign.
>
> An **equation** is a mathematical sentence, which often contains a variable.

There are times when you can figure out what the value of the variable has to be just by using your arithmetic facts or by a little guessing and testing. You can call that solving by inspection. That's fine when it works, but many times it's too difficult or too time consuming. You need a better strategy.

When you have an equation, something happened to the variable. Someone multiplied, or added, or did something (or several somethings) to the variable, and you know the result. When solving an equation, your job is to undo the arithmetic that has been performed and get the variable alone, or isolated, on one side of the equation. For example, if you start with the equation $3x - 2 = 25$, someone multiplied the variable by 3, then subtracted 2, and got 25. Your job is to undo that arithmetic and get to a simple "$x =$ the original number." In this case, $x = 9$.

Since you are undoing, you do the opposite of what has been done. The equation is like one of those old-fashioned scales, with a pan hanging on each side. The same amount of weight is on each pan, so the scale is balancing. To keep the equation balanced, you perform the same operation on both sides of the equation. Let's look at each of the basic steps first and then start combining them.

## Solving One-Step Equations

The simplest equations are the ones in which only one bit of arithmetic has been done to the variable. If I tell you that I picked a number and added 4 and my answer was 9, you can figure out what my number was by subtracting 4 from 9. If I pick a number, subtract 7 and get 15, you can find my number by adding. In the same way, you can undo multiplying by dividing, and undo division by multiplying. Let's look at some examples.

Solve the equation $x + 14 = 63$

Because 14 was added to the original value of the variable to get 63, you want to subtract 14 to get back to the original value. To keep the balance, subtract 14 from both sides. Subtracting 14 from the left side leaves just $x$, and subtracting 14 from 63 tells you the value of the variable is 49.

$$\begin{aligned} x + \cancel{14} &= 63 \\ \cancel{14} &= -14 \\ \hline x \quad\quad &= 49 \end{aligned}$$

To check your solution, write the original equation and replace the variable with the number you found. If the result is a true statement, your solution is correct. In this case, $x + 14 = 63$ is the original equation. You put your answer of 49 in place of $x$ and you have $49 + 14 = 63$, which is true. The solution $x = 49$ is correct.

Here's another example. This one asks you to undo subtraction.

Solve the equation $x - 117 = 238$

Someone took the number that $x$ stands for and subtracted 117. When they were done, there was 238 left. To get back to the original value of $x$, you need to add 117 to both sides of the equation. This will isolate $x$ and give you 355 as the value of $x$.

$$\begin{aligned} x - \cancel{117} &= 238 \\ \cancel{117} &= +117 \\ \hline x \quad\quad &= 355 \end{aligned}$$

You can check your solution by starting with $x - 117 = 238$ and putting 355 in place of $x$. Because $355 - 117 = 238$ is a true statement, you know your solution is correct.

Ready to try a multiplication equation? Solve the equation $19x = 28.5$

Undo the multiplying by dividing. Divide both sides of the equation by 19.

$$\begin{aligned} 19x &= 28.5 \\ \frac{\cancel{19}x}{\cancel{19}} &= \frac{28.5}{19} \\ x &= 1.5 \end{aligned}$$

Check your solution by replacing $x$ in the original equation with 1.5. $19x = 28.5$ becomes $19(1.5) = 28.5$ and that's true, so your solution is correct.

There are four operations of arithmetic so here's the fourth example. In this one, you see division, so use multiplication to solve it.

Solve the equation $\frac{x}{7} = 31$.

The variable was divided by 7 and the answer was 31. To find the value of the variable, multiply both sides by 7.

$$\frac{x}{7} = 31$$

$$\frac{x}{\cancel{7}} \cdot \cancel{7} = 31 \cdot 7$$

$$x = 217$$

Is 217 the correct value of $x$? Take the original equation and replace $x$ with 217. $\frac{x}{7} = 31$ becomes $\frac{217}{7} = 31$, which is true. The solution of 217 is correct.

With those four steps, you can tackle many different equations. You'll encounter equations that look much more complicated, but all that's necessary is to apply these simple steps one after another.

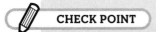
**CHECK POINT**

Solve each equation.

1. $x + 17 = 53$

2. $t - 11 = 46$

3. $-9a = 117$

4. $\frac{y}{6} = -14$

5. $x + 14 = -3$

## Solving Two-Step Equations

In each of the examples we've looked at, just one bit of arithmetic had been done to the variable, and so only one step was necessary to isolate the variable. More often, equations have more than one step. The key to solving two-step equations is performing the correct operations in the correct order.

Remember that the variable stands for a number. Imagine that you pick a number and then perform some arithmetic with that number. You would follow the order of operations, and you'd get an answer. The equation would show what you did and the answer you got. For example, if you picked a number, multiplied it by 7 and then added 5, and got an answer of 26, the equation would be $7x + 5 = 26$.

Solving an equation is undoing what was done by doing the opposite and stripping away the layers of arithmetic. To accomplish that, you need to undo things in the reverse of the order in which they were done. You undo the last bit of arithmetic first and work backward. Think about putting on your shoes and socks. You put a sock on first, then the shoe. But to take off your shoes and socks, you take the shoe off first, then the sock. It really wouldn't work in the other order.

To solve a two-step equation, take a moment to think about what was done to the variable. Notice what operations were performed and in what order. Then perform the opposite, or inverse, operations in the opposite order. So if you see that the variable was multiplied by 4 and then 9 was subtracted, you'll add 9 and then divide by 4. Let's look at some examples.

Solve $5x - 11 = 5$

Look at the equation and map out what happened to the variable. The number represented by $x$ was multiplied by 5 and then 11 was subtracted from the result. The answer was 5. So what happened was multiplication by 5, then subtraction of 11. The opposite of multiplication is division and the opposite of subtraction is addition, but you also want to reverse the order. Your plan for solving the equation will be first to add 11, then to divide by 5.

Add 11 to both sides.

$$5x \cancel{-11} = \quad 5$$
$$\cancel{11} = +11$$
$$5x \quad = \quad 16$$

Then divide both sides by 5.

$$5x \quad = \quad 16$$
$$\frac{\cancel{5}x}{\cancel{5}} = \frac{16}{5}$$
$$x = 3.2$$

You can check your solution of $x = 3.2$ by substituting 3.2 into the equation to see if it makes a true statement. $5(3.2) - 11 = 5$ becomes $16 - 11 = 5$, which is true, so your solution is correct.

Here's another example, with some different operations.

Solve $\frac{t}{4} + 3 = 7$

In this equation, $t$ is divided by 4, and then 3 is added, so the plan will be to subtract 3 and then multiply by 4. Take it one step at a time.

Subtract 3 from both sides of the equation.

$$\frac{t}{4} + 3 = \quad 7$$
$$-3 = -3$$
$$\frac{t}{4} \quad = \quad 4$$

Then multiply both sides of the equation by 4.

$$\frac{t}{4} \quad = \quad 4$$
$$\frac{t}{\cancel{4}} \cdot \cancel{4} = 4 \cdot 4$$
$$t = 16$$

Check the solution. Is $\frac{16}{4} + 3 = 7$ a true statement? It is, so the variable $t$ does represent 16.

Are you ready for another example? This one is a little different, just because the order of the operations is a little different.

Solve $\dfrac{y-7}{3} = -15$

In this case, the number represented by $y$ was reduced by 7—a subtraction was performed. Then that result was divided by 3. So your plan should be first to multiply by 3, then to add 7.

Multiplying both sides by 3 will eliminate the denominator.

$$\dfrac{y-7}{3} = -15$$

$$\dfrac{y-7}{\not{3}} \cdot \not{3} = -15 \cdot 3$$

$$y - 7 = -45$$

Adding 7 to both sides will isolate $y$ and tell you the value of the variable.

$$
\begin{array}{rl}
y - 7 &= -45 \\
+7 &= +7 \\
y &= -38
\end{array}
$$

Go back to the original equation to check your solution.

$\dfrac{y-7}{3} = -15$ becomes $\dfrac{-38-7}{3} = -15$ when you substitute -38 for $y$, and that is a true statement. The solution is correct.

---

**WORLDLY WISDOM**

A change in the order of things can sometimes make you feel confused about the solving process. An equation that's not in the order you're used to seeing, like the equation 5 - 3x = 2 can cause some trouble. When you look at what's happening in the equation, it's hard to know where to start and what happens in what order. Take a minute to rewrite the equation 5 - 3x = 2 as -3x + 5 = 2. It will be easier to plan your solution.

Let's look at one last example. This one can be solved two different ways. Let's look at one method here, and then we'll tackle the other method in the next section.

Solve -3(x − 1) = -27

In this case, someone took for a number represented by $x$ and subtracted 1 from it, then multiplied the result by -3. In the original problem, the parentheses tell you to do the subtraction first, then the multiplication. To solve the equation, you still want to do opposite operations in the opposite order, so divide by -3 and then add 1.

Undo the multiplication by dividing both sides by -3. When that's cleared, you won't need the parentheses anymore.

$$-3(x-1) = -27$$
$$\frac{\cancel{-3}(x-1)}{\cancel{-3}} = \frac{-27}{-3}$$
$$x-1 = 9$$

Then add 1 to both sides to find the value of $x$.

$$
\begin{array}{rcl}
x-1 &=& 9 \\
+1 &=& +1 \\
\hline
x &=& 10
\end{array}
$$

Check the solution. If $x$ is 10 and you subtract 1, you have 9, and 9 times -3 is -27. The solution of $x = 10$ is correct.

---

**CHECK POINT**

Solve each equation.

6. $9x - 7 = -43$

7. $\dfrac{y}{11} + 5 = 11$

8. $6 + 4x = 34$

9. $\dfrac{t-3}{5} = 12$

10. $7(x + 5) = 119$

## Variables on Both Sides

The basic technique for solving an equation is to do the opposite arithmetic operation to undo what's been done to the variable. When there are two steps, you undo them in the opposite order. Those rules are almost all you need to solve equations. The one piece that's left is how to deal with variables on both sides of the equation.

Until now, all the equations you were asked to solve had a single number, a constant, on one side. The other side had the variable and whatever was going on, and you knew you had to undo what was going on to get that variable all alone. But what if both sides of the equation had a variable? What if you had to solve $7x - 4 = 5x + 2$? You still need to isolate the variable, but which one?

It's not enough to get one of the $x$'s all alone. If you still have an $x$ on the other side, you won't know what number $x$ represents. The key to solving an equation with variables on both sides is to eliminate one variable term first. You can eliminate either variable term. The choice is yours.

In the equation $7x - 4 = 5x + 2$, you might choose to eliminate $5x$. You do that by subtracting $5x$ from both sides. Rules about like terms are important here. You can only subtract an $x$-term from another $x$-term. You'll subtract the $5x$ from the $7x$.

$$
\begin{array}{rl}
7x - 4 = & 5x + 2 \\
-5x\phantom{ - 4 } = & -5x \\
\hline
2x - 4 = & 2
\end{array}
$$

Once you've subtracted $5x$ from both sides, the second variable term is gone, and you have a two-step equation to solve.

$$
\begin{array}{rl}
2x - 4 = & 2 \\
+4 = & +4 \\
\hline
2x\phantom{ - 4 } = & 6 \\
\dfrac{\cancel{2}x}{\cancel{2}} = & \dfrac{6}{2} \\
x = & 3
\end{array}
$$

You can eliminate one variable term by adding or subtracting depending on the term you want to remove. In the equation $7 - 9x = 7x - 19$, you can eliminate the variable term on the right side by subtracting $7x$ from both sides.

$$
\begin{array}{rl}
7 - 9x = & 7x - 19 \\
-7x = & -7x \\
\hline
7 - 16x = & -19
\end{array}
$$

If you prefer, you can eliminate the variable term on the left side by adding $9x$ to both sides.

$$
\begin{array}{rl}
7 - 9x = & 7x - 19 \\
+9x = & +9x \\
\hline
7\phantom{ - 9x } = & 16x - 19
\end{array}
$$

Either way, you'll get the same solution.

If an equation has variable terms on both sides, eliminate one by adding or subtracting an equivalent variable term on both sides. Then solve the equation for the remaining variable.

## Simplifying Before You Solve

In an earlier example, I said there was another way to solve the equation $-3(x - 1) = -27$, and this is the moment to look at that method. When you solve equations, you want to be able to use those inverse, or opposite, operations, and having parentheses or extra terms can get in the way of that.

Before you begin the actual work of solving an equation, you want to make the equation as simple as possible. Focus on one side of the equation at a time, and if parentheses or other grouping symbols are present, remove them. You can do this by simplifying the expression inside the parentheses, by using the distributive property, or occasionally, by deciding that the parentheses are not necessary and just removing them. In the equation $-3(x - 1) = -27$, you can distribute the $-3$ and the equation will become $-3x + 3 = -27$. In the equation $(5x + 2) + (3x - 4) = 14$, the parentheses are really not necessary, so you can just drop them and the equation becomes $5x + 2 + 3x - 4 = 14$.

Once parentheses have been cleared, take the time to combine like terms (and *only* like terms) before you begin solving. Each side of the equation should have no more than one variable term and one constant term when you begin to solve. So in the equation $5x + 2 + 3x - 4 = 14$, you should combine the $5x$ and $3x$ and combine the $+2$ and the $-4$. The equation becomes $8x - 2 = 14$.

If there are variable terms on both sides of the equation, add or subtract to eliminate one of them. Next, add or subtract to eliminate the constant term that is on the same side as the variable term. You want to have one variable term equal to one constant term. Finally, divide both sides by the coefficient of the variable term.

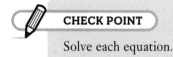

**CHECK POINT**

Solve each equation.

11. $11x + 18 = 3x - 14$        14. $4(5x + 3) + x = 6(x + 2)$

12. $5(x + 2) = 40$        15. $8(x - 4) - 16 = 10(x - 7)$

13. $5(x - 4) = 7(x - 6)$

## Special Cases

Sometimes when you try to isolate the variable, all the variable terms disappear. There are two reasons why this can happen. Sometimes you're working with an equation that will make a true statement no matter what value you substitute for the variable. The simplest example of this is the equation $x = x$. No matter what you replace $x$ with, you'll get a true statement. Equations like this are called *identities*. If you subtract $x$ from both sides of the equation, you find yourself with $0 = 0$, which is true, but not the "$x = $ a number" you were hoping for. In an identity, $x$ can equal any number. If all the variables disappear and what's left is true, you have an identity.

**DEFINITION**

An **identity** is an equation that is true for all real numbers.

The other reason why the variables may disappear is that you're working with an equation that is never true. The simplest example of this kind of equation is $x = x + 1$. There's no way a number could be equal to more than itself. In this case, if you subtract $x$ from both sides you get $0 = 1$, which is clearly not true. If all the variables disappear and what's left is false, the equation has no solution.

Don't confuse identities or equations with no solution with equations that have a solution of zero. The equation $0x = 0$ is an identity, the equation $0x = 4$ has no solution, but the equation $4x = 0$ has a solution of $x = 0$.

CHECK POINT

Solve each equation. If there is no solution, say that. If all numbers are solutions, label the equation an identity.

16. $2x - 3 = x - 3$

17. $6x - 4 = 2(3x - 2)$

18. $9x + 11 = 3(3x + 4)$

19. $5x + 13 = 22 - 4x$

20. $19 - 8x = 4(5 - 2x) - 1$

# One Solution or Many?

When you solve an equation, you find the value of the variable that makes the two sides of the equation identical. It's the one number that makes the statement true. There are more complicated equations for which there is more than one solution. We're not ready to look at the solution of those equations just yet, but it's good to know they exist. For example, the equation $x^2 = 9$ has two solutions. If you replace $x$ with 3, you have $3^2 = 9$, which is true, but if you replace $x$ with -3, you get $(-3)^2 = 9$, which is also true. The equation $x^2 = 9$ has two solutions.

There's another group of problems that are true for more than one value of the variable, and these are a type of problem we are ready to solve. They're called *inequalities* because rather than saying "this equals that," they say, "this is bigger than that," or "this is smaller than that." They tell us that the two sides are unequal.

DEFINITION

An **inequality** is a mathematical sentence that says one expression is greater than another.

The symbol $>$ is read "is greater than," and the symbol $<$ is read "is less than." If you add a line under the symbol, you add "or equal to." So $x \le 4$ says "$x$ is less than or equal to 4," and $y \le -3$ says "$y$ is less than or equal to -3. If you think about those statements for a minute, you can see that there are many numbers that you could put in place of $x$ to make a true statement. You could say $5 \ge 4$ or $19 \ge 4$ and both would be true. There are many, many more substitutions for $x$ that make true statements—an infinite number of them, in fact.

Equations ask you to find the value for the variable that makes both sides the same. Inequalities ask you to find the values that make one side larger than the other. The solution will be a set of numbers, rather than a single value.

## Solving Inequalities

Inequalities can be solved in much the same way as equations, with one important exception. When you multiply or divide both sides of an inequality by a negative number, the direction of the inequality sign reverses. Remember that the positive and negative sides of the number line are mirror images of one another. When you multiply both sides of the inequality by a negative number, you go through the looking glass and things change. On the positive side, 5 is bigger than 4, but flip to the negative side and -5 is smaller than -4.

The rules for solving inequalities are the same as those for solving equations, except for that one step. When you multiply or divide both sides of an inequality by the coefficient of the variable term, you have to make a decision about the inequality sign.

If you divide both sides of an inequality by a positive number, leave the inequality sign as is.

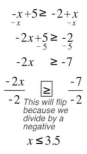

$$3x - 5 \geq 22$$
$$\phantom{3x} +5 \quad +5$$
$$3x \geq 27$$
$$\frac{3x}{3} \geq \frac{27}{3}$$
$$x \geq 9$$

If you divide both sides of an inequality by a negative number, reverse the inequality sign.

$$-x + 5 \geq -2 + x$$
$$-x \qquad -x$$
$$-2x + 5 \geq -2$$
$$\phantom{-2x} -5 \quad -5$$
$$-2x \geq -7$$
$$\frac{-2x}{-2} \;\boxed{\geq}\; \frac{-7}{-2}$$

This will flip because we divide by a negative

$$x \leq 3.5$$

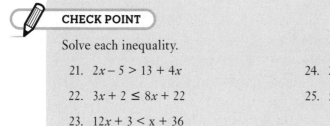

**CHECK POINT**

Solve each inequality.

21. $2x - 5 > 13 + 4x$

22. $3x + 2 \leq 8x + 22$

23. $12x + 3 < x + 36$

24. $2y - 13 \geq 4(2 + y)$

25. $5x - 10(x - 1) > 95$

## Picturing the Solution

The solution of an equation is usually just one number, so when you say $x = 5$, it's easy to understand what that means. The solution of an inequality is a set of numbers, a collection that goes on and on and includes whole numbers, rational numbers, and irrational numbers. When you write $x > 5$, you're describing a whole collection of numbers. It's helpful to have a picture to understand it.

The solution set of an inequality can be graphed on the number line by shading the appropriate portion of the line. You can graph the inequality $x \geq 2$ on the number line by putting a solid dot on 2 and then shading all the numbers to the right of 2.

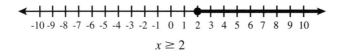

$$x \geq 2$$

Start the graph of the inequality $x > -4$ by circling -4. Don't fill in that circle, because you don't actually want -4, but you want everything less than -4. Then shade to the left.

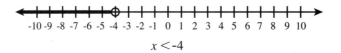

$$x < -4$$

Use an open circle if the inequality sign is $>$ or $<$ and a solid dot for inequalities containing $\leq$ or $\geq$.

**CHECK POINT**

Graph the solution set of each inequality.

26.  $x > 6$                     29.  $a \leq 2$

27.  $t < -1$                    30.  $x > 0$

28.  $y \geq -3$

## The Least You Need to Know

- Solve equations by performing the opposite operations in the opposite order.
- Always simplify before trying to solve.
- In an equation, a variable usually represents just one number. In an inequality, a variable usually represents a set of numbers.
- Solve inequalities just like equations, but if you multiply or divide both sides by a negative number, reverse the inequality sign.

# Coordinate Graphing

When you learned to solve equations and inequalities, I mentioned that there should only be one variable involved. If there are two variables in the same equation, the solution would have to be not just a number, but a pair of numbers, one for each variable. It turns out that if you have two variables and only one equation, there are infinitely many different pairs of numbers that could be solutions. You can't settle on just one pair. If you look at an equation like $x + y = 5$, you could say $x$ is 1 and $y$ is 4, or $x$ is 2 and $y$ is 3, or $x$ is 5 and $y$ is 0. That's three possible solutions already, and we haven't begun to talk about negative numbers or fractions yet. A single equation with two variables has many, many possible solutions.

When you learned to solve an inequality, you learned that a picture of the solution set on a number line was helpful in understanding what the solution really meant. In this chapter, you'll learn a system for picturing the many, many solutions for an equation with two variables. You'll look at the basic idea of the system and at quick ways to draw the picture for your particular equation. You'll also see how a few key pieces of information from the picture tell you what equation it represents. Inequalities with two variables can be pictured too, and you'll see how those pictures compare to the pictures of equations.

## In This Chapter

- How to plot points on a graph
- How to graph lines quickly
- Finding the slope of a line
- Graphing linear equations and inequalities

# The Coordinate Plane

Take a sheet of paper and draw a horizontal number line across the middle of the sheet. Then draw a vertical number line, so that the two lines have their zero in the same spot. With just those two lines in place, you can direct someone to any point on the paper by giving them a number on the horizontal line and one on the vertical line. It's as if those two numbers were the names of streets, 4ᵗʰ Street and 5ᵗʰ Avenue, and you wanted to meet someone on the corner.

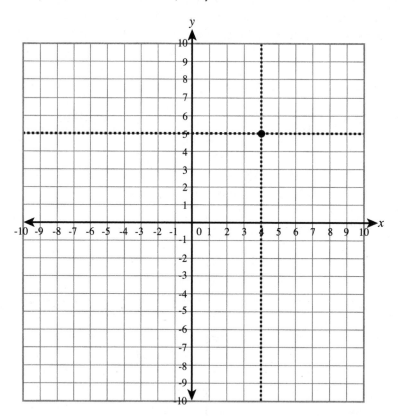

The horizontal number line is called the $x$-axis, and the vertical one is called the $y$-axis. Every point is represented by a pair of numbers, $(x,y)$. The point in our example is the point $(4,5)$. The two numbers are called the *coordinates* of the point. 4 is the $x$-coordinate, and 5 is the $y$-coordinate. To locate the point, start at the spot that is 0 on both number lines. This point $(0,0)$ is called the origin. Count left or right according to the first coordinate. In this case, count 4 to the right. Then let the $y$-coordinate tell you how far up or down to go. In this case, go up 5.

> **DEFINITION**
>
> The Cartesian coordinate system, named for René Descartes, is a rectangular coordinate system that locates every point in the plane with an ordered pair of numbers, (*x,y*). The **x-coordinate** indicates horizontal movement, and the **y-coordinate** vertical movement.

The *x*-axis and *y*-axis divide the graphing area into four sections called quadrants. The following graph shows the point (3,7) in the upper right quadrant, which is quadrant I. The point (-2,5) is in the upper left quadrant, quadrant II. In the lower left quadrant, called quadrant III, you can see the point (-3,-1), and in quadrant IV on the lower right, the point (4,-3). The point (5,0) sits on the *x*-axis, and (0, 4) is on the *y*-axis.

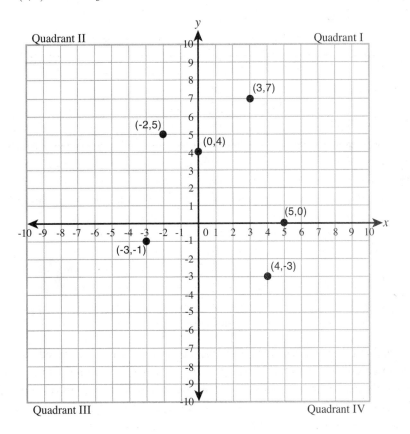

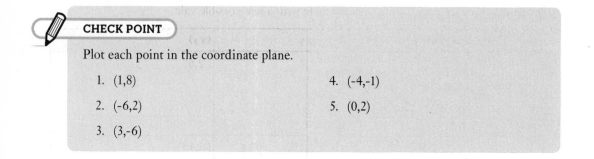

**CHECK POINT**

Plot each point in the coordinate plane.

1. (1,8)                              4. (-4,-1)
2. (-6,2)                             5. (0,2)
3. (3,-6)

# Graphing Linear Equations

Our coordinate system assigns a pair of numbers to every point and a point to every pair of numbers. The power of such a system is that it lets you give a picture of all the pairs of numbers that solve an equation or inequality with two variables.

## Graphs as Pictures of Patterns

The graph of an equation in two variables is a picture of all the pairs of numbers that balance the equation. An equation in two variables has infinitely many solutions, each of which is an ordered pair $(x,y)$. The graph of the equation is a picture of all the possible solutions. Because each of those pairs of numbers fits the same rule, when you plot the points, you find that they fall in a pattern, specifically a line. That's why these equations are sometimes called linear equations.

## Plotting Points

The most straightforward way to graph an equation is to choose several values for $x$, substitute each value into the equation, and calculate the corresponding values for $y$. This information can be organized into a table of values. Two points are technically enough to determine a line, but when building a table of values, it is wise to include several more, so that any errors in arithmetic will stand out as deviations from the pattern.

To graph the equation $3x + 2y = 6$, make a table with a few possible values of $x$.

| $x$ | $3x + 2y = 6$ | $y$ | $(x,y)$ |
|---|---|---|---|
| -2 | $3(-2)+ 2y = 6$<br>$-6 + 2y = 6$<br>$2y = 12$<br>$y = 6$ | 6 | $(-2,6)$ |
| -1 | $3(-1) + 2y = 6$<br>$-3 + 2y = 6$<br>$2y = 9$<br>$y = 4.5$ | 4.5 | $(-1,4.5)$ |
| 0 | $3(0) + 2y = 6$<br>$0 + 2y = 6$<br>$2y = 6$<br>$y = 3$ | 3 | $(0,3)$ |
| 1 | $3(1) + 2y = 6$<br>$3 + 2y = 6$<br>$2y = 3$<br>$y = 1.5$ | 1.5 | $(1,1.5)$ |
| 2 | $3(2) + 2y = 6$<br>$6 + 2y = 6$<br>$2y = 0$<br>$y = 0$ | 0 | $(2,0)$ |

Once you've built your table, plot each of your points. You should see them falling into a line. If a point doesn't fall in line, check your arithmetic.

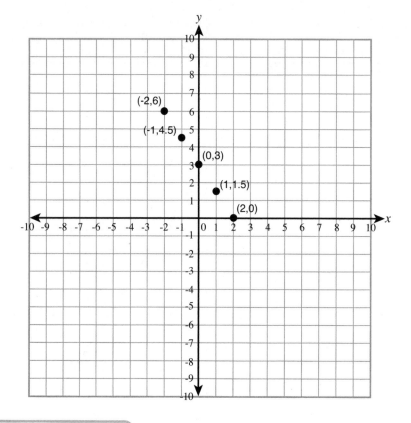

If the coefficient of $x$ is a fraction, choose $x$-values that are divisible by the denominator of the fraction. This will minimize the number of fractional coordinates, which are hard to estimate.

Once you have a line of points, connect them and extend the lines in both directions. Add an arrow to each end to show that the line continues.

When you build a table of values, make a habit of choosing both positive and negative values for $x$. Of course, you can choose $x = 0$, too. Usually, you'll want to keep the $x$-values near zero so that the numbers you're working with don't get too large. If they do, you'll need to extend your axes, or re-label your scales by twos or fives, or whatever multiple is convenient.

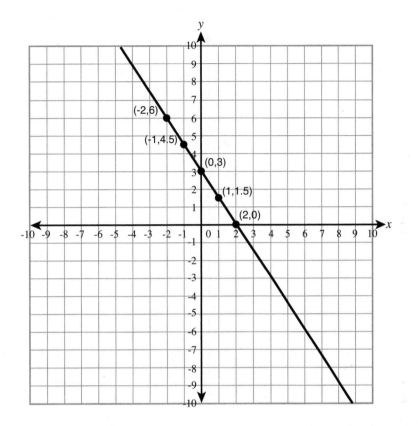

**CHECK POINT**

Make a table of values and graph each equation.

6. $x + y = 7$

7. $2x - y = 3$

8. $y = 3x - 6$

9. $y = 8 - 2x$

10. $y = \dfrac{2}{3}x - 1$

## Quick Graphing

Making a table and plotting points will always get you a graph, but it can be a slow process. There are two ways to get the graph quickly, and it's good to know both, because the way the equation is arranged will determine which method works better.

The first method uses the fact that a point on an axis will always have one coordinate that's 0. Points on the $x$-axis have 0 as their $y$-coordinate, and points on the $y$-axis have 0 as their $x$-coordinate. These points on the $x$-axis and $y$-axis are called intercepts, and the method is called the intercept-intercept method.

To graph $3x - 4y = 12$ by the intercept-intercept method, replace $x$ with 0 and find $y$. $3(0) - 4y = 12$ becomes just $-4y = 12$, so $y = -3$. The point $(0,-3)$ is the $y$-intercept. Go back to the original equation and let $y = 0$. $3x - 4(0) = 12$ becomes just $3x = 12$ and $x = 4$. The $x$-intercept is $(4,0)$. Plot the $x$-intercept and the $y$-intercept, connect them, and extend to make the line $3x - 4y = 12$.

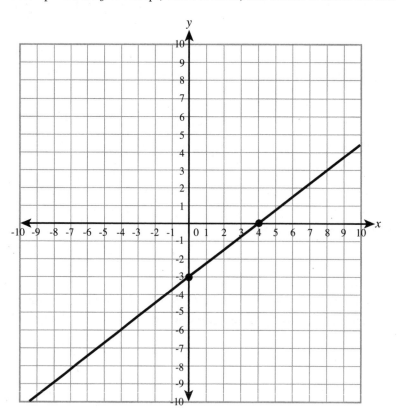

Although it can be used for graphing any equation, the intercept-intercept method is best used when the equation has the $x$ and $y$ terms on the same side and the constant on the other side. When the equation is arranged that way, the arithmetic of finding the intercepts is usually simple.

**CHECK POINT**

Graph each equation by intercept-intercept.

11.  $x + y = 10$

12.  $6x + 2y = 12$

13.  $2x - 3y = 9$

14.  $x - 2y = 8$

15.  $6x + 2y = 18$

The other quick graphing method uses the $y$-intercept and a pattern we notice in lines, called the *slope* of the line. The slope of a line is a measurement of the rate at which the line rises or falls. A rising line has a positive slope, and a falling line has a negative slope. A horizontal line has a slope of zero, and a vertical line has an undefined slope. A line with a slope of 4 is steeper than a line with a slope of 3. A line with a slope of -3 falls more steeply than a line with a slope of -1.

**DEFINITION**

The **slope** of a line is a number that compares the rise or fall of a line to its horizontal movement.

The slope of a line is found by counting from one point on the line to another and making a ratio of the up or down motion to the left or right motion. The up or down motion is called the rise, and the left or right motion is called the run. So the slope is $\frac{\text{rise}}{\text{run}}$.

The traditional symbol for the slope is $m$. If you know two points on the line, you can find the rise by subtracting the $y$-coordinates and the run by subtracting the $x$-coordinates. If the points are $(x_1, y_1)$ and $(x_2, y_2)$, then

$$m = \frac{\text{rise}}{\text{run}} = \frac{y_2 - y_1}{x_2 - x_1}.$$

Because you find the slope by subtracting the $y$'s to find the rise and subtracting the $x$'s to find the run, the formula for the slope can be written as $m = \frac{y_2 - y_1}{x_2 - x_1}$.

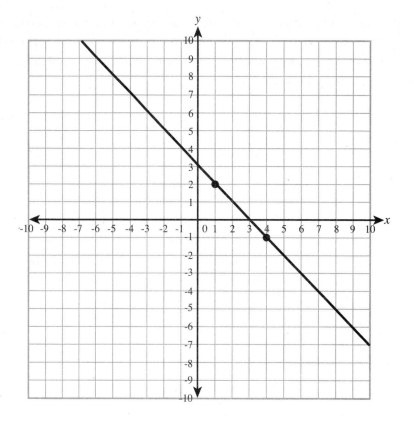

The slope of the line through the points (4,-1) and (1,2) is

$$m = \frac{2 - \text{-}1}{1 - 4} = \frac{3}{-3} = \text{-}1$$

**CHECK POINT**

Find the slope of the line that connects the given points.

16.  (7,2) and (4,5)                    19.  (-5,5) and (5,-1)

17.  (6,-4) and (9,-6)                  20.  (3,4) and (8,4)

18.  (4,6) and (8,7)

To draw the graph of an equation quickly, arrange the equation so that $y$ is isolated and the $x$ term and constant term are on the other side. This is called $y = mx + b$ form. The value of $b$ is the $y$-intercept of the line, and the value of $m$ is the slope of the line. In the equation $y = \frac{1}{2}x - 5$, the $y$-intercept is (0,-5) and the slope is $\frac{1}{2}$.

Begin by plotting the $y$-intercept, then count the rise and run and plot another point. Repeat a few times and connect the points to form a line. For the equation $y = \frac{1}{2}x - 5$, start by plotting the $y$-intercept at -5 on the $y$-axis, then count up 1 (the rise) and 2 to the right (the run), and place a point. Use the slope to plot a few more points by counting up 1 and over 2 a few more times, putting a dot each time. If any points seem out of line, double check your count. Connect the points into a line and extend it in both directions.

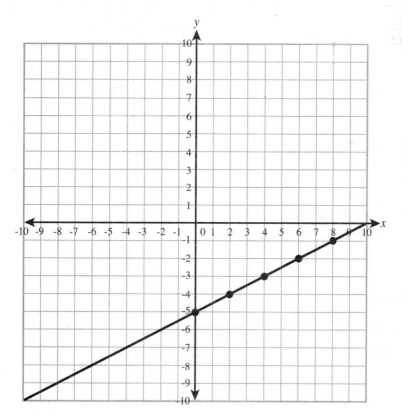

**CHECK POINT**

Graph each equation using $y$-intercept and slope.

21. $y = -\dfrac{3}{4}x + 1$

22. $y = -4x + 6$

23. $y = -3x - 4$

24. $2y = 5x - 6$

25. $y - 6 = 3x + 1$

## Vertical and Horizontal Lines

Horizontal lines fit the $y = mx + b$ pattern, but since they have a slope of zero, they become $y = 0$. Whatever value you may choose for $x$, the $y$-coordinate will be $b$.

Vertical lines have undefined slopes, so they cannot fit the $y = mx + b$ pattern, but since every point on a vertical line has the same $x$-coordinate, they can be represented by an equation of the form $x = c$, where $c$ is a constant. The value of $c$ is the $x$-intercept of the line.

**CHECK POINT**

Graph each equation.

26. $y = -3$

27. $x = 2$

28. $y = 5$

29. $x = -1$

30. $y + 1 = 4$

# Graphs of Inequalities

The graph of an equation is a picture of all the pairs of numbers that solve the equation, so the graph of an inequality should be the picture of all the pairs of numbers that make the inequality true. That may (or may not) include points on a line, but it will include a lot of other points, either above or below the line.

To graph an inequality like $y < \dfrac{3}{4}x - 2$, begin by graphing the line $y = \dfrac{3}{4}x - 2$. If the inequality sign is $\geq$ or $\leq$ , you would use a solid line, but here, for $>$ or $<$, use a dotted line.

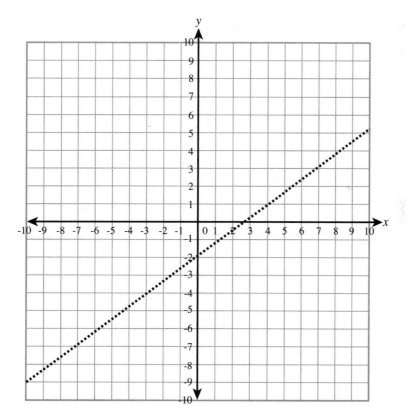

You want the points with $y$-coordinates that are less than $\frac{3}{4}x - 2$. The points on the line have $y$-coordinates equal to that, one side of the line has points with $y$-coordinates that are greater and the other side of the line has the ones that are less. Test a point on one side of the line in the inequality; the origin is often a convenient choice. If the result is true, shade that side of the line; if not, shade the other side.

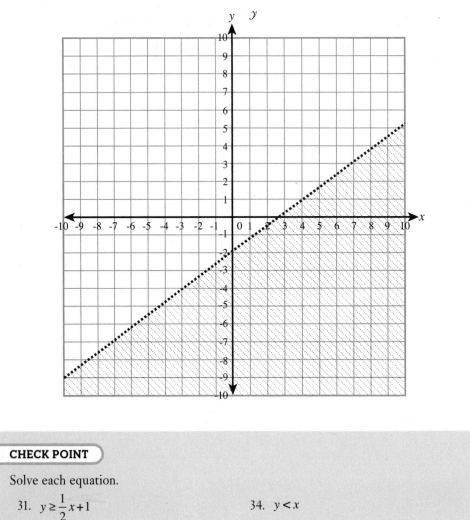

---

✏️ **CHECK POINT**

Solve each equation.

31. $y \ge \frac{1}{2}x + 1$

32. $y \le 2x - 5$

33. $y > 5x - 4$

34. $y < x$

35. $y \le -\frac{2}{3}x + 5$

## The Least You Need to Know

- The coordinate system assigns a pair of numbers, called coordinates, to every point in the plane. The first number tells you how to move horizontally, and the second tells how to move vertically.
- Any equation can be graphed by making a table and plotting points. The x-intercept and the y-intercept are easy points to calculate.
- The slope is a number that tells whether a line rises or falls and how steeply.
- If the equation is in $y = mx + b$ form, start at $b$ on the y-axis and count the slope, $m$, to see where the line goes.

# PART 3

# The Shape of the World

Any world tour must include seeing the sights, having a look at the shape of things. Whether it's a natural formation like a mountain or a canyon, or a famous building or monument, by the end of a journey, your scrapbook will certainly have some examples of great geometry.

This part of our journey is devoted to looking at the geometry of the mathematical world. You'll learn the basic vocabulary you need to describe what you see and how to identify the polygons, circles, polyhedrals, and other solids that form the architecture of the mathematical world.

We won't abandon numbers completely as we look at these shapes, but for a while numbers will be a lesser focus. There's still a place for calculating, but we'll investigate relationships between shapes as well.

# CHAPTER 12

# Basics of Geometry

Geometry begins with a few undefined terms—point, line, and plane—and logically builds a system that describes many physical objects. The word "geometry" means "earth measuring," and geometry had its beginnings in the work of dividing land up among farmers. Doing that requires lines, angles, and many different shapes.

In this chapter, you'll lay the foundation on which to build your knowledge of geometry. Starting from those undefined terms, you'll learn about portions of lines and combining lines to make angles. You'll measure and classify angles, explore relationships between them, and bisect segments and angles. Parallel and perpendicular lines are the building blocks of many figures and create many angle relationships. Last of all, you'll take many of these ideas onto the coordinate plane and see how they connect back to algebra.

## In This Chapter

- The vocabulary of geometry
- Exploring angle relationships
- Parallel and perpendicular lines
- Basic geometry and the coordinate plane

# Points, Lines, Planes, and Angles

The undefined terms of geometry are a curious mix of things you know and things you can only imagine. A *point* can be thought of as a dot, a tiny spot, or a position. That's the familiar part. The part that requires imagination is the idea that a point doesn't take up any space. It has no size and no dimension. You can't measure it. You can draw a dot to represent a point, even though your dot does take up some space, and you label points with uppercase letters.

You know what a line is. You see lines all the time. But geometry asks you to use your imagination here, too. A *line* is a set of points—an infinite string of points—that goes on forever in both directions. It has length, in fact it has infinite length, but it has no width and no height. It's only one point wide or high, and points don't take up space. And yet, somehow, you can string points together to make something that has infinite length.

People often say, "a straight line," but in geometry that phrase is redundant. All lines are straight. If it curves or bends, it's not a line.

When you draw a picture to represent a line, even though your picture does have some width and can't actually go on forever, you put arrows on the ends to show that it keeps going. You can label the line with one script letter, like line $\ell$, or by placing two points on the line and writing those two points with a line over the top, like this: $AB$.

> **DEFINITION**
>
> A **point** is a position in space that has no length, width, or height. A **line** is a set of points that has length but no width or height. A **plane** is a flat surface that has length and width but no thickness. **Space** is the set of all points.

Is your imagination still working? Where do these points and lines live? Where would you draw a point or a line? Perhaps on a sheet of paper or the chalkboard? Those surfaces are the images that help you imagine a plane. A *plane* is a flat surface that has infinite length and infinite width but no height or thickness. It's an endless sheet of paper that's only one point deep.

And where do the points and the lines and the planes live? In space! No, not outer space, at least not exactly. *Space*, in geometry, is the set of all points, everywhere.

# Basic Geometry Terms

Before your imagination is completely exhausted, let's look at some parts of a line that don't require so much imagination. A *ray*, sometimes called a half-line, is a portion of a line from one point, called the endpoint, going on forever in one direction. You can't measure the length of a ray because, like a line, it goes on forever. A ray looks like an arrow, and you name it by naming its endpoint and then another point on the ray, with an arrow over the top, like this: $\overrightarrow{AB}$.

More familiar, if only because it can really fit on your paper is a line segment, literally, part of a line. A *line* is a portion of a line between two endpoints. (Finally, something you can measure!) Name it by its endpoints, with a segment over the top, like this: $\overline{AB}$.

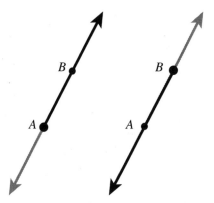

Lines contain infinitely many points, but they are named by any two points on the line. A line that contains the two points A and B can be named $\overleftrightarrow{AB}$ or $\overleftrightarrow{BA}$. In the same way, a line segment can be named by its endpoints in either order, but for rays, the order makes a difference. The rays $\overrightarrow{AB}$ and $\overrightarrow{BA}$ are shown and are two different rays.

 **DEFINITION**

> A **ray** is a portion of a line from one endpoint, going on forever through another point.
>
> A **line segment** is a point of a line made up of two endpoints and all the points of the line between the endpoints.

When you put two rays together, you create a new figure called an angle. An *angle* is a figure formed by two rays with a common endpoint, called the *vertex*. The two rays are the *sides* of the angle. You'll often see angles whose sides are line segments, but you can think of those segments as parts of rays. (By the way, you can measure angles, too.)

**DEFINITION**

> An **angle** is two rays with a common endpoint, called the **vertex**. The rays are the **sides** of the angle.

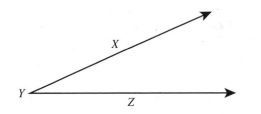

In the angle $\angle XYZ$, the vertex is $Y$, and the sides are rays $\overrightarrow{YX}$ and $\overrightarrow{YZ}$. You can name angles by three letters, one on one side, the vertex, and one on the other side. $\angle XYZ$ and $\angle ZYX$ are both names for this angle. An angle can be named by just its vertex, for example, $\angle Y$, as long as it is the only angle with that vertex.

---

**CHECK POINT**

Draw and label each figure described.

1. Line segment $\overline{PQ}$

2. Ray $\overrightarrow{YZ}$

3. Angle $\angle DEF$

4. Rays $\overrightarrow{AB}$ and $\overrightarrow{AC}$

5. Angles $\angle PQR$ and $\angle RQT$

## Length and Angle Measure

Well, you found things you can measure: a line segment or an angle. They don't go on forever. You can measure them and actually stop somewhere. So how do you do it?

Measuring just means assigning a number to something to give an indication of its size. The number depends on the ruler you're using. Feet? Inches? Centimeters? Furlongs? They all measure length (although you don't see furlongs used much outside of horse racing).

A *ruler* is just a line or line segment that you've broken up into smaller segments, all the same size, and numbered. You could even use a number line, and many times we will.

If you place a ruler next to a line segment, each endpoint of the segment will line up with some number on the ruler (even if it's one of the little fraction lines in between the whole numbers).

The numbers that correspond to the endpoints are called *coordinates*, and the length of the line segment is the difference between the coordinates. Technically, the length is the absolute value of the difference, because direction doesn't matter.

**DEFINITION**

A **ruler** is a line or segment divided into sections of equal size, labeled with numbers, called **coordinates**, used to measure the length of a line segment.

*A number line like this can be used to measure line segments. For example, the length of line segment AB is equal to the distance between coordinates -7 and -2, or five units.*

You can also measure angles. Angles are measured by the amount of rotation from one side to the other. Picture the hands of a clock rotating, creating angles of different sizes. It is important to remember that the lengths of the sides have no effect on the measurement of the angle. The hands of the famous clock known as Big Ben are much longer than the hands of your wrist watch or alarm clock, but they all make the same angle at 9 o'clock.

So how do you put a ruler on an angle? For starters, it's not a ruler. A ruler is a line you use to measure parts of lines. Angles aren't parts of lines. They're more like wedges from a circle. So to measure them you create an instrument called a *protractor*, a circle broken into 360 little sections, each called a degree.

In geometry, angles are measured in degrees. When you put the protractor over the angle with the center of the circle on the vertex of the angle, the sides fall on numbers, called coordinates. The measure of the angle is the absolute value of the difference of the coordinates.

### DEFINITION

A **protractor** is a circle whose circumference is divided into 360 units, called degrees, which is used to measure angles.

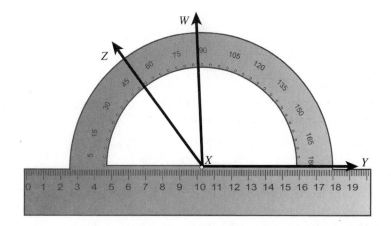

*A protractor can be created using any circle, but most people are familiar with the plastic half-circle tool shown here.*

When two segments have the same length, they are called *congruent* segments. In symbols, you could write $\overline{AB} \cong \overline{XY}$ to say that the segment connecting $A$ to $B$ is the same length as the segment connecting $X$ to $Y$. You could also write $AB = XY$ to say the measurements—the lengths—are the same. With the little segment above the letters, you're talking about the segment. Without it, you're talking about the length, a number. Segments are congruent. Lengths are equal.

> **DEFINITION**
>
> Two segments are **congruent** if they are the same length. Two angles are **congruent** if they have the same measure.

The same is true of angles and their measures. The symbol $\angle A$ refers to the actual angle, and the symbol $m\angle A$ denotes the measure of that angle. If you write $\angle XYZ \cong \angle RST$, you're saying the two angles have the same measure. You could also write $m\angle XYZ = m\angle RST$. Angles are congruent; measures are equal.

A full rotation all the way around the circle is 360°. Half of that, or 180°, is the measure of a *straight angle.* The straight angle takes its name from the fact that it looks like a line.

An angle of 90°, or a quarter rotation, is called a *right angle.* If one side of a right angle is on the floor, the other side stands upright. Angles between 0° and 90° are called *acute angles.*

Angles whose measurement in greater than 90° but less that 180° are *obtuse angles.*

> **DEFINITION**
>
> A **straight angle** is an angle that measures 180°. A **right angle** is an angle that measures 90°.
>
> An angle that measures less than 90° is an **acute angle**. An **obtuse angle** is an angle that measures more than 90° but less than 180°.

You can classify angles one by one, according to their size, but you can also label angles based on their relationship to one another. Sometimes the relationship is about position or location or what the angles look like. Other times it's just about measurements.

Two angles whose measurements total to 90° are called *complementary angles.* If two angles are complementary, each is the complement of the other.

The complement of an angle of 25° can be found by subtracting the known angle, 25°, from 90°. $90° - 25° = 65°$, so an angle of 25° and an angle of 65° are complementary. To find the measure of the complement of an angle of 12°, subtract $90° - 12° = 78°$. An angle of 12° and an angle of 78° are complementary angles.

Two angles whose measurements total to 180° are called *supplementary angles*. If two angles are supplementary, each is the supplement of the other.

📖 **DEFINITION**

> **Complementary angles** are a pair of angles whose measurements total 90°.
>
> **Supplementary angles** are a pair of angles whose measurements total 180°.

To find the supplement of an angle of 132°, 180° − 132° = 48°. The measure of the supplement of an angle of 103° is 180 − 103 = 77°.

When two lines intersect, the lines make an X and four angles are formed. Each pair of angles across the X from one another is a pair of *vertical angles*. Vertical angles are always *congruent;* they always have the same measurement.

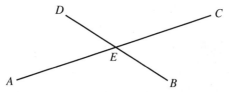

*The angles ∠AED and ∠CEB are vertical angles, as are the angles ∠DEC and ∠AEB.*

Two angles that have the same vertex and share a side but don't overlap are called *adjacent angles*. Two adjacent angles whose exterior sides (the ones they don't share) make a line are called a *linear pair.* Linear pairs are always supplementary.

📖 **DEFINITION**

> **Vertical angles** are a pair of angles both of which have their vertices at the point where two lines intersect and do not share a side.
>
> **Adjacent angles** are a pair of angles that have the same vertex and share a side but do not overlap one another.
>
> A **linear pair** is made up of two adjacent angles whose unshared sides form a straight angle.

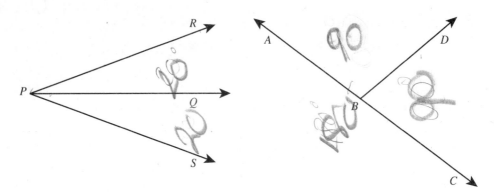

*Angles ∠RPQ and ∠SPQ are adjacent. Angles ∠ABD and ∠CBD are both adjacent and linear.*

---

**CHECK POINT**

6.  If m∠X = 174°, then ∠X is a(n) ⟨obtuse⟩ angle.

7.  If m∠T = 38°, then ∠T is a(n) _____.

8.  If ∠X and ∠Y are supplementary, and m∠X = 174°, then m∠Y = _____.

9.  If ∠R and ∠T are complementary, and m∠T = 38°, then m∠R = _____.

10. Lines $\overrightarrow{PA}$ and $\overrightarrow{RT}$ intersect at point Y. If m∠PYR = 51°, then m∠RYA = _____ and m∠TYA = _____.

---

## Midpoints and Bisectors

While you're in the middle of all these lines and segments and rays and angles, it's a good time to talk about middles. Because lines and rays go on forever, you can't talk about the middle of a line or the middle of a ray. To say where the middle of something is, you have to be able to measure it. Until you can assign a length to an object, you can't say where halfway is.

A *midpoint* is a point on a line segment that divides it into two segments of equal length, two congruent segments. If $M$ is the midpoint of $\overline{AB}$, then $\overline{AM} \cong \overline{MB}$. Each of the little pieces is the same length ($AM = MB$), and each of them is half as long as $\overline{AB}$. Only segments have midpoints.

A line or ray or segment that passes through the midpoint of a segment is a *segment bisector.*

Angles don't have midpoints, but they can have bisectors. An *angle bisector* is a ray from the vertex of the angle that divides the angle into two congruent angles.

> 📖 **DEFINITION**
>
> The **midpoint** of a line segment is a point on the segment that divides it into two segments of equal length.
>
> A **segment bisector** is a line, ray, or segment that divides a segment into two congruent segments.
>
> An **angle bisector** is a line, ray, or segment that passes through the vertex of an angle and cuts it into two angles of equal size.

> ✏️ **CHECK POINT**
>
> 11. $M$ is the midpoint of segment $\overline{PQ}$. If $PM = 3$ cm, $MQ =$ _____ cm and $PQ =$ ___10___ cm.
>
> 12. $H$ is the midpoint of $\overline{XY}$. If $XY = 28$ inches, then $XH =$ _____ inches.
>
> 13. Ray $\overrightarrow{AB}$ bisects $\angle CAT$. If $m\angle CAT = 86°$, then $m\angle HAT =$ _____.
>
> 14. If $m\angle AXB = 27°$ and $m\angle BXC = 27°$, then _____ bisects $\angle AXC$.
>
> 15. $m\angle PYQ = 13°$, $m\angle QYR = 12°$, $m\angle RYS = 5°$, and $m\angle SYT = 20°$. True or False: $\overleftrightarrow{YR}$ bisects $\angle PYT$.

# Parallel and Perpendicular Lines

*Perpendicular lines* are lines that intersect at right angles. The symbol for "is perpendicular to" is ⊥, so you can write $\overrightarrow{XY} \perp \overleftrightarrow{AB}$ to say that segment $\overline{XY}$ is perpendicular to line $\overleftrightarrow{AB}$. If a line, ray, or segment is perpendicular to another segment and also divides that segment into two congruent parts, that line, ray, or segment is the *perpendicular bisector* of the segment.

When two lines are perpendicular, all the angles at the intersection will be right angles. When two lines intersect, the vertical angles are congruent, and the adjacent angles are linear pairs, so if one of the angles is a right angle, all four will be right angles.

Perpendicular lines intersect to form right angles, but what about lines that don't intersect? Lines in the same plane that are always the same distance apart and therefore never intersect are called *parallel lines*. You see parallel lines all the time: the edges of windows and doors, the lines on a sheet of notebook paper, and railroad tracks are all parallel.

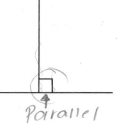
Parallel

> **DEFINITION**
>
> Lines, rays or segments that meet to form a right angle are **perpendicular**.
>
> **Parallel lines** are lines on the same plane that never intersect.

Parallel lines all by themselves are not all that interesting. They just keep on going, never meet, and don't do anything exciting. When other lines get mixed up with parallel lines, however, some more interesting things do happen.

## The Defining Angles

When parallel lines are cut by another line, called a *transversal*, eight angles are formed. Different pairs from this group of eight are classified in different ways.

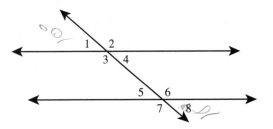

As the transversal crosses the first line, it creates a cluster of four angles, labeled ∠1, ∠2, ∠3, and ∠4 in this picture. As it crosses the second line, it creates another cluster of four angles, labeled ∠5, ∠6, ∠7, and ∠8. In each cluster, there is an angle in the upper left position (∠1 from the top cluster or ∠5 from the bottom). There are also angles in the upper right, lower left, and lower right positions. The angle from the upper cluster and the angle from the lower cluster that are in the same position are called *corresponding angles*.

> **DEFINITION**
>
> A **transversal** is a line that intersects two or more other lines.
>
> **Corresponding angles** are a pair of angles created when a transversal intersects two parallel lines that are on the same side of the transversal and are both above or both below the parallel lines.

When parallel lines are cut by a transversal, corresponding angles are congruent. They have the same measurements. ∠1 ≅ ∠5, ∠2 ≅ ∠6, ∠3 ≅ ∠7, and ∠4 ≅ ∠8.

Look at only the angles that are between the parallel lines, ∠3, ∠4, ∠5, and ∠6. Choose one from the top cluster, say ∠3, and the one from the bottom cluster on the other side of the transversal, ∠ 6, and you have a pair of *alternate interior angles.* Take one angle from each side of the transversal so that they are not between the parallels but outside them, like ∠1 and ∠8 or ∠2 and ∠7, and you have a pair of *alternate exterior angles.* Alternate exterior angles are congruent if the lines are parallel.

 **DEFINITION**

> **Alternate interior angles** are two angles formed when a transversal intersects parallel lines that are on opposite sides of the transversal and between the parallel lines.
>
> **Alternate exterior angles** are two angles formed when a transversal intersects parallel lines that are on opposite sides of the transversal and outside the parallel lines.

When parallel lines are cut by a transversal, alternate interior angles are congruent. They have the same measurements. ∠3 ≅ ∠6 and ∠4 ≅ ∠5.

Using these facts about parallel lines and corresponding angles and alternate interior angles, and the fact that vertical angles are congruent, you can figure out that ∠1 ≅ ∠4 ≅ ∠5 ≅ ∠8 and ∠2 ≅ ∠3 ≅ ∠6 ≅ ∠7. Add the fact that ∠1 and ∠2 are supplementary, and it becomes possible to assign each of the angles one of two measurements. ∠1, ∠4, ∠5, and ∠8 have one measurement, and ∠2, ∠3, ∠6, and ∠7 are supplementary to those. If you know the measurement of one angle created when a transversal cuts parallel lines, you can find the measurements of all eight angles. If m∠1 is 40°, then ∠4, ∠5, and ∠8 are also 40°, and the other four angles are 180° − 40° = 140°.

**CHECK POINT**

Lines $\overleftrightarrow{PQ}$ and $\overleftrightarrow{RT}$ are parallel. Transversal $\overleftrightarrow{AB}$ intersects $\overleftrightarrow{PQ}$ at $X$ and $\overleftrightarrow{RT}$ at $Y$.

16.  ∠*PXY* and ∠*XYT* are a pair of _____ angles.

17.  ∠*AXQ* and ∠*XYT* are a pair of _____ angles.

18.  If m∠*XYT* = 68°, then m∠*PXA* = _____.

19.  If m∠*PXY* = 107°, then m∠*RYB* = _____.

20.  If $\overleftrightarrow{AB} \perp \overleftrightarrow{PQ}$, then m∠*XYR* = _____.

## Slopes

Geometry and algebra may feel like different worlds at times, but now and then, they come together, and that often happens on the coordinate plane. When we looked at graphing linear equations, you saw how the slope of a line controlled its tilt or angle. When you talk about parallel and perpendicular lines, the angles the lines make are important.

The graphs of two lines in the coordinate plane will be parallel lines if they have the same slope. The matching slopes mean they run at the same angle and don't tilt toward each other, so they never cross. The line $y = 2x - 3$ and the line $y = 2x + 5$ both have a slope of 2, and so they will be parallel.

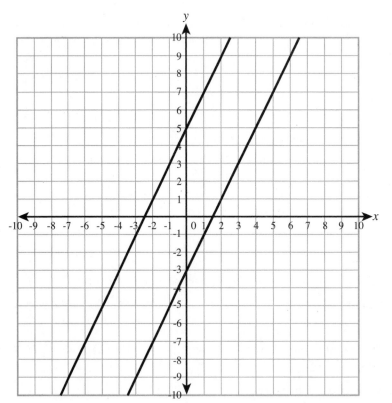

In order for the graphs of two linear equations to be perpendicular lines, one must rise and one must fall, so the slopes must have opposite signs. That alone won't get that exact right angle, however. To actually be perpendicular, the lines must have slopes that are negative reciprocals. If one line has a slope of 2, a line perpendicular to it will have a slope of $-\frac{1}{2}$. The graphs of $y = -\frac{3}{5}x + 4$ and $y = \frac{5}{3}x - 1$ are perpendicular lines because their slopes, $-\frac{3}{5}$ and $\frac{5}{3}$, multiply to -1, so they are negative reciprocals.

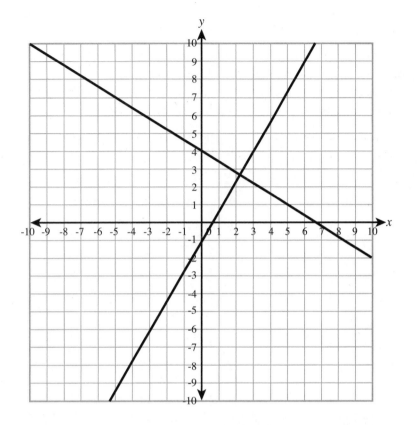

**CHECK POINT**

Label each pair of lines as parallel, perpendicular, or neither.

21. Line $a$ has a $y$-intercept of 2 and a slope of -3. Line $b$ has a $y$-intercept of $-\frac{1}{2}$ and a slope of -3.

22. Line $\overleftrightarrow{RT}$ has a slope of $\frac{3}{5}$ and a $y$-intercept of -4. Line $\overleftrightarrow{PQ}$ has a slope of $-\frac{5}{3}$ and a $y$-intercept of 2.

23. The equation of line $p$ is $y = 2x - 5$, and the equation of line $q$ is $y = 5x - 2$.

24. Line $\overleftrightarrow{XY}$ and line $\overleftrightarrow{WZ}$ if $X$ is the point (4,-2), $Y$ is (7,6), $W$ is (2,-8), and $Z$ is (0, 4).

25. The line $3x - 2y = 12$ and the line $2x + 3y = 12$.

## The Least You Need to Know

- Midpoints and bisectors divide segments into two congruent segments; angle bisectors divide angles into two congruent angles.
- Angles less than 90° are acute, angles greater than 90° are obtuse. Right angles are 90°, and straight angles are 180°.
- Complementary angles are two angles that add to 90°, and supplementary angles are two angles that add to 180°.
- Vertical angles are congruent; linear pairs are supplementary.
- Parallel lines never intersect; perpendicular lines intersect at right angles.
- If parallel lines are cut by a transversal, pairs of corresponding angles, alternate interior angles, or alternate exterior angles are congruent.

# CHAPTER 13

# Triangles

Geometry is a branch of mathematics that looks at shapes. That's true, but "shapes" covers so much ground that the statement really doesn't tell you much. To understand what geometry really looks at, what kinds of questions it tries to answer, you need to break that down a bit.

The first big distinction you can make is that geometry looks at shapes with straight sides, called polygons, and shapes that curve, or circles. There are some things you can say about all polygons, but the conversation gets more interesting when you start to break that group down even more according to the number of sides the polygon has.

In this chapter, we'll explore triangles, polygons with three sides. We'll look at some things that are true about all triangles, and then organize triangles into groups, based on facts about their sides or about their angles, or both. There will be lots to learn about right triangles, the most interesting group. We'll also have a look at the very practical matters of area and perimeter.

## In This Chapter

- Classifying triangles by sides and angles
- How to find missing angles and measure exterior angles
- The Pythagorean theorem and how to use it
- How to find area and perimeter of triangles

# Facts about Triangles

A *polygon* is a figure formed by connecting line segments at their endpoints until they come back to where they started. You need at least three segments to be able to close the shape, but you can make a polygon with many more sides. The line segments are called sides, and the points where the sides meet are called vertices. When two sides meet at a vertex, they form an angle.

**DEFINITION**

A **polygon** is a closed figure made up of line segments that meet at their endpoints.

The members of the family of polygons take more specific names based on the number of sides they have. A polygon with three sides is called a triangle. It has three sides, three vertices, and three angles. The prefix *tri* means "three." Tri-angle = three angles. A polygon with four sides is called a quadrilateral. *Quad* means "four" and *lateral* means "side." We'll look at quadrilaterals in the next chapter.

Math is always interested in measuring things, and geometry is no exception. In a triangle, you can measure the sides and the angles. You can also draw and measure other line segments inside the triangle.

An *altitude* is a line segment that starts at a vertex of the triangle and meets the opposite side at a right angle. You measure the altitude to find the height of the triangle, so often the words altitude and height are used interchangeably.

**DEFINITION**

An **altitude** is a line segment from a vertex of a triangle perpendicular to the opposite side.

The altitude usually falls inside the triangle, but sometimes it can actually be a side of the triangle, and sometimes, if the triangle has an obtuse angle, the altitude can be outside the triangle. If that happens, you extend the opposite side to see where the altitude should stop.

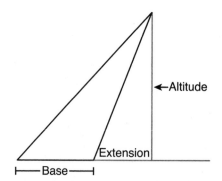

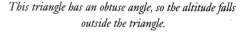

*This triangle has an obtuse angle, so the altitude falls outside the triangle.*

You can bisect any one of the angles in a triangle. The line segment from the vertex to the opposite side that cuts the angle into two congruent angles will not necessarily cut the opposite side in half. It's an *angle bisector*, not a segment bisector.

> **DEFINITION**
>
> An **angle bisector** in a triangle is a line segment from a vertex to the opposite side that divides the angle at that vertex into two congruent angles.

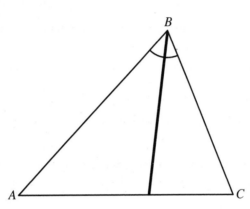

*The segment drawn from angle B to side AC bisects angle B by cutting it into two angles of equal size.*

The line segment from a vertex to the midpoint of the opposite side is called a *median*.

> **DEFINITION**
>
> A **median** of a triangle is a line segment that connects a vertex to the midpoint of the opposite side.

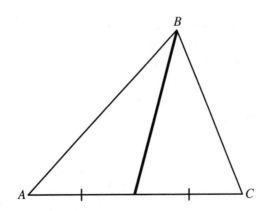

*The segment drawn from angle B to the midpoint of side AC divides AC into two equal segments but does not bisect angle B.*

In most triangles, the altitude, the angle bisector, and the median from a vertex are all different line segments.

It's often the case that something is true only for some triangles, but there are things that are true about any triangle. One important one is called the Triangle Sum theorem. In any triangle, the sum of the measures of the three angles is 180°. You can see this if you cut a triangle out of paper and tear off two corners and set them next to the third corner.

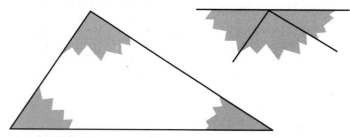

*The torn corners can be put together to form a 180° angle.*

If $\triangle ABC$ is a triangle with $\angle B = 90°$, the sum of the measures of $\angle A$ and $\angle C$ must be 90° because the three angles of the triangle total 180°. If $\angle B$ measures 90°, the other two angles must make up the other 90°.

An *exterior angle* of a triangle is formed by extending one side of the triangle. The measure of an exterior angle of a triangle is equal to the sum of the two remote interior angles.

> 📖 **DEFINITION**
>
> If one side of a triangle is extended, the adjacent angle formed is called an **exterior angle** of the triangle.

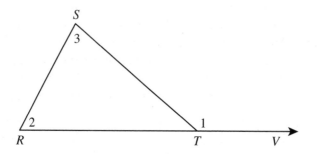

*The sum of the two remote interior angles (m∠1 and m∠2) equals the measure of the exterior angle (m∠3):*
*m∠1 + m∠2 = m∠3.*

In $\triangle RST$, m$\angle S = 43°$ and m$\angle R = 28°$. The measure of the exterior angle of the triangle at $T$ is equal to the sum of the two remote interior angles, $\angle S$ and $\angle R$. Adding the measures of those two angles tells you the exterior angle is $43 + 28 = 71°$. You could calculate the measure of $\angle STR$ ($180 - 43 - 28 = 109°$), and since $\angle STV$ is supplementary to $\angle STR$, it will be $180 - 109 = 71°$.

So for any triangle, you know what the angles add up to, but you don't know what the individual angles measure. When you look at sides, you can again say something that's true all the time, but there's a limit to how specific you can be. In any triangle, the sum of the lengths of any two sides will be greater than the length of the third. Imagine your home is at one vertex, $A$, of a triangle, and your best friend's home is at another vertex, $C$. To get to your friend's home, you could walk down the street from $A$ to $B$ and then turn the corner and follow the road from $B$ to $C$. Or you could take the shortcut through the backyards straight from $A$ to $C$. The shortcut is, obviously, shorter. $AB + BC > AC$.

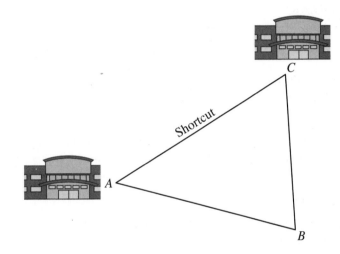

Put another way, the length of any side of a triangle is less than the sum of the other two sides. Suppose you know a triangle has sides 4 inches long and 7 inches long. Those add to 11 inches. The third side must be less than 11 inches. But turn things around a little bit. The unknown side plus the 4-inch side must be more than 7 inches, so the unknown side must be more than 3 inches. It has to be more than 3 but less than 11 inches. The length of any side of a triangle is less than the sum of the other two sides but more than the difference between them.

The longest side of a triangle is always opposite the largest angle. Remember that angles are measured by the rotation between the sides. The wider an angle opens, the longer the line segment it will take to connect the ends of the sides. The longest side is opposite the largest angle, and the shortest side is opposite the smallest angle. That, of course, means that the medium-sized side is opposite the medium-sized angle. It also means that whenever a triangle has congruent angles, the sides opposite the congruent angles are the same length.

1. In △*RST*, m∠*R* = 48°, and m∠*T* = 102°. Find the measure of ∠*S*.

2. In △*RST* (described in question 1), side $\overline{RS}$ is extended through *S* to point *Q* to create exterior angle ∠*TSQ*. Find the measure of ∠*TSQ*.

3. In △*RST* (described in question 1), side $\overline{RS}$ is extended through *S* to point *Q* to create exterior angle ∠*TSQ* and through *R* to point *P* to create exterior angle ∠*TRP*, and side $\overline{RT}$ is extended through *T* to point *N* to form exterior angle ∠*STN*. Find the total of m∠*TSQ* + m∠*TRP* + m∠*STN*.

4. Placidville is 43 miles from Aurora, and Aurora is 37 miles from Lake Grove. The distance from Placidville to Lake Grove is at least _____ miles but less than _____ miles.

5. Gretchen lives 5 miles from the library and 2 miles from school. The distance from the library to school is between _____ miles and _____ miles.

# Classifying Triangles

Just as trying to talk about all polygons at the same time didn't make much sense, trying to talk about all triangles at once can be hard. It's more practical to break the family of triangles down into categories, and you usually do that either by the sides or by the angles.

If all the sides of a triangle have different lengths, the triangle is *scalene*. You can't say much else about them, except that all their angles will have different measures, too.

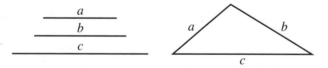

*A **scalene triangle** has three sides of different lengths.*

*Isosceles triangles* are triangles with two congruent sides. The two congruent sides are called the legs, and the third side is called the base. The angles opposite the congruent sides, often called the *base angles*, are congruent to each other. The angle between the legs is usually called the *vertex angle*.

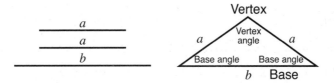

*An **isosceles triangle** has two sides that are the same length. The congruent sides are called the **legs**, and the third side is the **base**. The angle between the equal sides is the **vertex angle**, and the other two angles are **base angles**.*

In an isosceles triangle, the altitude drawn from the vertex to the base bisects the base and the vertex angle. This is one of the few cases where one line segment does more than one job. It's an altitude, it's an angle bisector, and it's a median.

An equilateral triangle is one in which all three sides are the same length. All three angles will be the same size, and because the three angles add up to 180°, each of the angles in an equilateral triangle measures 60°. Any altitude is a super segment that bisects the side to which it is drawn and the angle from which it is drawn.

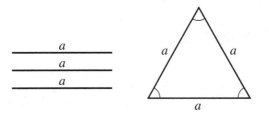

*A triangle in which all three sides are the same length is an **equilateral triangle**. Equilateral triangles are also equiangular.*

Because the three angles of an equilateral triangle are congruent, it is also described as *equiangular.* All three angles in an equiangular triangle measure 60°.

> ### 💡 WORLDLY WISDOM
>
> Is an equilateral triangle isosceles? It depends on who you talk to. Some folks will say an isosceles triangle has *exactly* two equal sides, so they would say no. Others will say an isosceles triangle has at least two equal sides, and their answer would be yes. The fact is, anything that's true for an isosceles triangle is true for an equilateral triangle.

The other ways that triangles are classified by angles have to do with the sizes of the angles. If all three angles are acute angles, that is, if all three have measurements less than 90°, then you say the triangle is an *acute triangle.* This can happen in a scalene triangle if, for example, it has angles of 80°, 47°, and 53°, or it can happen in an equilateral triangle that has angles of 60°, 60°, and 60°. An isosceles triangle can be acute, too. For example, you might have angles of 30°, 75°, and 75°.

If a triangle has an obtuse angle, an angle greater than 90°, it's called an obtuse triangle. The altitude of an obtuse triangle will often fall outside the triangle.

If a triangle has a right angle, it's called a right triangle. We'll take a closer look at right triangles in the next section.

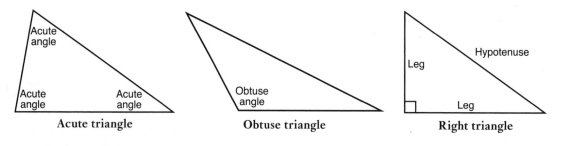

**Acute triangle**          **Obtuse triangle**          **Right triangle**

> 💡 **WORLDLY WISDOM**
>
> Obtuse triangles can only have one obtuse angle, and right triangles can only have one right angle. The three angles of a triangle add up to 180°. If one angle is already 90° or more, there's 90° or less for the other two. They'll have to be acute.

✏️ **CHECK POINT**

6. $\triangle RST$ is isosceles with $RS = ST$. If m$\angle SRT = 39°$, then m$\angle STR = \underline{39°}$.

7. In right triangle $\triangle ABC$, $\overline{AB} \perp \overline{BC}$ and m$\angle C$ is 19°. Find m$\angle A$.

8. True or False: If m$\angle P = 17°$ and m$\angle Q = 25°$, then $\triangle PQR$ is an acute triangle.

9. If m$\angle P = 17°$ and m$\angle Q = 25°$, then the longest side of $\triangle PQR$ is side _____.

10. If the vertex angle of an isosceles triangle measures 94°, then the base angles measure _____.

# Right Triangles

Right triangles are triangles that contain one right angle, and they pop up all over the place. When someone wants to build something that stands up vertically, they'll often add a slanting support to create a right triangle. Every time you draw an altitude in any triangle, you create two right triangles. The right triangle is the one in which a side can also be an altitude.

In a right triangle, the two sides that meet to form the right angle are called the legs, and the third side, opposite the right angle, is called the hypotenuse. The hypotenuse of a right triangle is always the longest side because it's opposite the right angle, the largest angle of the triangle.

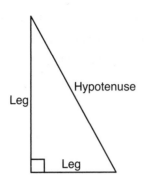

*In a right triangle, the two sides that form the right angle are called **legs**, and the side opposite the right angle is called the **hypotenuse**.*

## The Pythagorean Theorem

One of the most famous theorems in mathematics can be applied to right triangles. It's named for Pythagoras, a sixth century B.C.E. Greek mathematician and philosopher. What Pythagoras actually said about right triangles was probably something like "the square constructed on the hypotenuse of a right triangle contains the squares on the other two sides." His method of investigating was actually drawing squares, but you can think about it from more of a number point of view.

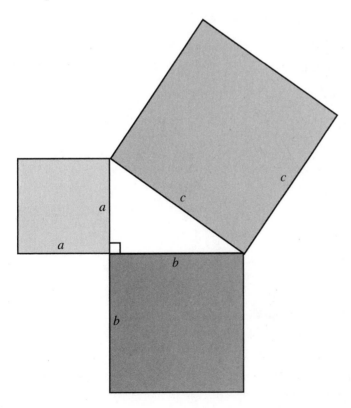

In any right triangle, if you measure all the sides and square those measurements, the square of the length of the hypotenuse will be equal to the sum of the squares of the other two sides. If the legs of a right triangle measure 5 inches and 12 inches, then the hypotenuse is 13 inches; $13^2 = 169$ and $5^2 + 12^2 = 25 + 144 = 169$.

The easiest way to remember the Pythagorean theorem, and the most common way, is in symbolic form. If the legs of the right triangle are labeled $a$ and $b$ and the hypotenuse is $c$, then $a^2 + b^2 = c^2$.

> 📖 **DEFINITION**
>
> The Pythagorean theorem states that for any right triangle, the square of the hypotenuse is equal to the sum of the squares of the other two sides, or if the length of the hypotenuse is $c$, and the lengths of the legs are $a$ and $b$, then $a^2 + b^2 = c^2$.

The importance of the Pythagorean theorem comes from the fact that you can use it to find the length of one side of the right triangle if you know the other two.

Suppose $\triangle RST$ is a right triangle with right angle at $R$. If $RS = 7$ and $RT = 4$, you can find the length of side $\overline{TS}$ with the Pythagorean theorem. The right angle is at $R$, so the sides you know are the legs that make the right angle, and you're looking for the hypotenuse. Use $a^2 + b^2 = c^2$, with $a = 7$ and $b = 4$.

$c^2 = 49 + 16$

$c^2 = 65$

This means TS will be $\sqrt{65}$, which is slightly more than 8.06.

You've probably thought about problems that could be solved with the Pythagorean theorem. Here's one. Elise walks to school every morning, and on sunny days, she can cut across the football field from corner to corner. On snowy days, she must go around the outside edges of the field. How much shorter is Elise's walk on sunny days?

The field has a right angle at its corner, and Elise's path from corner to corner makes the hypotenuse of a right triangle. The field, including the end zones, is 120 yards long and 53.33 yards wide. Use $a^2 + b^2 = c^2$, with $a = 120$ and $b = 53.33$ to find the length of Elise's shortcut.

$c^2 = a^2 + b^2$

$c^2 = (120)^2 + (53.33)^2$

$c^2 = 14,400 + 2,844.0889$

$c^2 = 17,244.0889$

$c = \sqrt{17,244.0889}$

$c \approx 131.32$

On sunny days, Elise can take the shortcut that's about 131.32 yards, but on snowy mornings, she'll have to go around the edges of the field, a total of $120 + 53.33 = 173.33$ yards. On sunny days, she saves $173.33 - 131.32 \approx 42.01$ yards. Her sunny day route is about 42 yards shorter.

You can use the Pythagorean theorem to find a leg, too, if you know one leg and the hypotenuse. Suppose you're shopping for a new TV, and you're looking at one that is advertised as 55 inch. That's the diagonal measurement, corner to corner, so it's the hypotenuse of a right triangle. If that television is 27 inches high, how wide will it be? The height is one leg, and the width you're looking for is the other leg, so $a = 27$, $b$ is unknown, and $c = 55$.

$$a^2 + b^2 = c^2$$
$$27^2 + b^2 = 55^2$$
$$729 + b^2 = 3{,}025$$
$$b^2 = 3{,}025 - 729 = 2{,}296$$
$$b = \sqrt{2{,}296} \approx 47.9$$

The TV will be about 48 inches, or 4 feet, wide.

You've probably notices a lot of "approximately equal to" answers from the Pythagorean theorem. That's because a lot of those square roots produce irrational numbers, whose decimals go on forever and have to be rounded.

There are some problems that work out to nice whole number answers, and when you work with right triangles, you get to know them. A set of three whole numbers that fits the Pythagorean theorem is called a *Pythagorean triple*. The most common one is 3-4-5: $3^2 + 4^2 = 5^2$. Multiples of Pythagorean triples are also triples, so 6-8-10 and 30-40-50 work as well. Other Pythagorean triples are 5-12-13 and 8-15-17. These sets of numbers come up a lot in right triangle problems, and recognizing the triples can save you some work.

> **DEFINITION**
>
> A **Pythagorean triple** is a set of three whole numbers $a$, $b$, and $c$ that fit the rule $a^2 + b^2 = c^2$.

> **CHECK POINT**
>
> Find the missing side of each right triangle.
>
> 11. In $\triangle XYZ$, $\overline{XY} \perp \overline{YZ}$. If $XY = 15$ cm and $YZ = 20$ cm, find $XZ$.
>
> 12. In $\triangle RST$, $\overline{ST} \perp \overline{RT}$. If $ST = 20$ inches and $RS = 52$ inches, find $RT$.
>
> 13. In $\triangle PQR$, $\overline{PQ} \perp \overline{PR}$. If $PQ = PR = 3$ feet, find $QR$.
>
> 14. In $\triangle CAT$, $\overline{CA} \perp \overline{AT}$. If $CT = 8$ meters and $CA = 4$ meters, find $AT$.
>
> 15. In $\triangle DOG$, $\overline{DO} \perp \overline{OG}$. If $DO = 21$ cm and $DG = 35$ cm, find $OG$.

## Special Right Triangles

The Pythagorean triples show up a lot when you're working with right triangles, and certain families of right triangles tend to show up a lot as well. When an altitude is drawn in an equilateral triangle, it divides the triangle into two right triangles. Do you remember that this altitude is one of those super segments that are altitudes, angle bisectors, and medians all in one? Each of these smaller triangles has a right angle where the altitude meets the base, an angle of 30° where the altitude bisects the vertex angle, and an angle of 60°. These right triangles are often called 30-60-90 right triangles because of their angles.

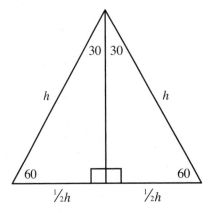

The hypotenuse of the 30°-60°-90° triangle is the side of the original equilateral triangle. The side opposite the 30° angle is half as large because that altitude was also a median, so it divided that side into two congruent segments. Using the Pythagorean theorem, you can find the length of the third side of the right triangle, the side that actually is the altitude. Suppose the hypotenuse is 1 foot long. That means the leg you know is half of that, or $\frac{1}{2}$ foot.

$$a^2 + b^2 = c^2$$

$$\left(\frac{1}{2}\right)^2 + b^2 = 1^2$$

$$\frac{1}{4} + b^2 = 1$$

$$b^2 = \frac{3}{4}$$

$$b = \sqrt{\frac{3}{4}} = \frac{\sqrt{3}}{\sqrt{4}} = \frac{\sqrt{3}}{2} = \frac{1}{2}\sqrt{3}$$

The side opposite the 60° angle must be half the hypotenuse times the square root of 3. So if you know the side of an equilateral triangle in 8 inches long, the altitude will divide the base into two segments, each 4 inches long, and the altitude itself will be $4\sqrt{3}$ inches long, or approximately 6.9 inches.

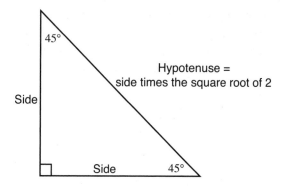

The other special right triangle, another one that shows up a lot, is the 45-45-90 triangle, or the isosceles right triangle. In an isosceles right triangle, the two legs are of equal length. You could pick your favorite number for an example, but let's say the legs are each 6 inches long. Apply the Pythagorean theorem. The equation $a^2 + b^2 = c^2$ becomes $6^2 + 6^2 = c^2$, or $c^2 = 72$. Take the square root and $c$, the length of the hypotenuse, is $c = \sqrt{72} = 6\sqrt{2}$. The hypotenuse of an isosceles right triangle is equal to the length of the leg times the square root of 2.

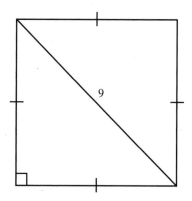

Suppose the diagonal of a square is 9 cm. The diagonal of the square is the hypotenuse of an isosceles right triangle, so if that is equal to 9, and you know it's supposed to be equal to the side times the square root of 2, you can say $s\sqrt{2} = 9$ and $s = \dfrac{9}{\sqrt{2}} \approx 6.4$.

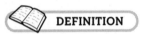

**CHECK POINT**

16. $\triangle ABC$ is a 30°-60°-90° right triangle, with hypotenuse 8 cm long. Find the length of the shorter leg.

17. $\triangle RST$ is an isosceles right triangle with legs 5 inches long. Find the length of the hypotenuse.

18. $\triangle ARM$ is a right triangle with $AR = 14$ meters, $RM = 28$ meters, and $AM = 14\sqrt{3}$ meters. Find the m$\angle M$.

19. $\triangle LEG$ is a right triangle with $LE = EG$ and $LG = 7\sqrt{2}$ inches. Find m$\angle G$.

20. $\triangle OWL$ is an isosceles right triangle with $OW > OL$. Which angle is the right angle?

# Area and Perimeter

The *perimeter* of any figure is the distance around all the edges. In a triangle, that means the sum of the lengths of the three sides. If a right triangle has legs of 3 feet and 4 feet and a hypotenuse of 5 feet, its perimeter is $3 + 4 + 5 = 12$ feet.

**DEFINITION**

The **perimeter** of a triangle (or any polygon) is the total of the lengths of all its sides.

If you know the perimeter and two of the sides of a triangle, you can work backward to find the length of the other side. If a triangle has a perimeter of 42 inches, and you know that it has an 18-inch side and a 15-inch side, you can add $18 + 15$ to find out that the two known sides account for 33 inches of the perimeter, so the third side must be $42 − 33 = 9$ inches long.

If you think of the perimeter as the outline of a shape, you can think of the area as the space inside the perimeter. This is true for triangles and for all polygons. When you talk about the size of a lot of land or the size of a rug, you're talking about area.

**DEFINITION**

The **area** of a polygon is the space enclosed within its sides.

To find the area of a triangle, you need to know the measure of the base and the height. Then you can plug these measurements into a simple formula, $A = \frac{1}{2}bh$, where $A$ = area, $b$ = base, and $h$ = height. You can call any one of the sides the base, as long as the height is the length of the altitude drawn from the opposite vertex, perpendicular to the base. This can sometimes cause the altitude to fall outside the triangle. If that happens, extend the side to cross the altitude. The length of the altitude is from the vertex to the point where it crosses the extension, but the length of the base is only the part in the triangle. It doesn't include the extension.

Suppose you want to find the area of an equilateral triangle with sides 12 inches long. You need to find the length of an altitude, but if you remember the special right triangles, it's not too bad. The altitude in an equilateral triangle is also a median and an angle bisector, so it creates two $30°$-$60°$-$90°$ triangles. The length of the altitude is half the length of a side times the square root of three, so $6\sqrt{3}$ inches. Use the side of 12 inches as the base and the $6\sqrt{3}$ inches as the height.

$$A = \frac{1}{2}bh$$

$$A = \frac{1}{2}(12)\left(6\sqrt{3}\right)$$

$$A = 36\sqrt{3} \approx 62.35$$

The area of the triangle is approximately 62.35 square inches.

**WORLDLY WISDOM**

Area is always measured in square units: square inches, square feet, square centimeters, etc. When you multiply feet times feet you get square feet. Meters times meters yields square meters.

Triangle *PQR* has an area of 24 square centimeters. If the lengths of its sides are 3 centimeters, 6 centimeters, and 8 centimeters, find the length of the longest altitude.

The area of the triangle will be the same no matter which side is called the base, if the altitude is drawn to that base. If we say the base is the 3 centimeter side, then $A = \frac{1}{2}bh$ becomes $24 = \frac{1}{2} \cdot 3 \cdot h$ and $h = 16$. If we use the 6 centimeter side as the base, then $A = \frac{1}{2}bh$ becomes $24 = \frac{1}{2} \cdot 6 \cdot h$ and $h = 8$. Declare that the base is the 8 centimeter side, then $A = \frac{1}{2}bh$ becomes $24 = \frac{1}{2} \cdot 8 \cdot h$ and $h = 6$. The longest altitude is 16 cm.

**CHECK POINT**

21. Find the area of a triangle with a base 14 cm long and an altitude of 7 cm.

22. If the area of a triangle is 27 square inches, and the altitude measures 6 inches, how long is the base to which that altitude is drawn?

23. Find the perimeter of a right triangle with legs that measure 20 cm and 48 cm.

24. Find the perimeter of an equilateral triangle with an area of $9\sqrt{3}$ square inches.

25. The area of a right triangle with legs of 3 cm and 4 cm and hypotenuse of 5 cm is _____ square centimeters, so the altitude from the right angle to the hypotenuse is _____ centimeters long.

## The Least You Need to Know

- Equilateral triangles have three sides of equal length, isosceles triangles have two sides of equal length, and scalene triangles have no equal sides.

- Obtuse triangles contain one obtuse angle, right triangles contain one right angle, and acute triangles have three acute angles.

- In any right triangle, the sum of the squares of the lengths of the legs is equal to the square of the length of the hypotenuse.

- In a 30-60-90 right triangle, the shorter leg is half the length of the hypotenuse and the longer leg is half the hypotenuse times the square root of three.

- The perimeter of any polygon is the sum of the lengths of its sides.

- The area of a triangle is one-half the length of the base times the height.

# CHAPTER 14

# Quadrilaterals and Other Polygons

As you work your way deeper into geometry, you'll begin to encounter more complex figures. We started with just lines, then angles, and in the last chapter, we worked with triangles. In this chapter we'll begin looking at polygons—shapes with more sides and angles.

The primary focus in this chapter is on four-sided figures, and we'll look at several different subgroups of the family of four-sided polygons. For each of the families, we'll consider the special properties of sides, angles, and diagonals that belong to that family. After you've gotten to know all the members of the family, we'll look at finding their perimeters and areas. Before moving on, we'll explore polygons with even more sides and learn some interesting facts about them.

## In This Chapter

- The properties of parallelograms
- Identifying special types of parallelograms
- The properties of trapezoids
- How to find the area and perimeter of quadrilaterals
- Polygons with more than four sides

## Parallelograms

The term *quadrilateral* is used for any four-sided polygon, but most of the attention falls on the members of the family called parallelograms. The name *parallelogram* comes from the fact that these quadrilaterals are formed by parallel line segments.

A parallelogram is a quadrilateral with two pairs of opposite sides parallel. Whenever you look at one pair of parallel sides, the other sides can be thought of as transversals. You know a bit about the angles that are formed when parallel lines are cut by a transversal, and you can see that you have consecutive interior angles that are supplementary.

**DEFINITION**

A **quadrilateral** is a polygon with four sides. A **parallelogram** is a quadrilateral in which both pairs of opposite sides are parallel and congruent.

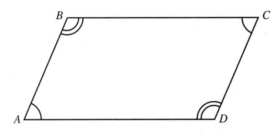

In parallelogram *ABCD*, and in any parallelogram, consecutive angles are supplementary. In *ABCD*, that means m∠*A* + m∠*B* =180°, m∠*B* + m∠*C* =180°, m∠*C* + m∠*D* =180°, and m∠*D* + m∠*A* =180°. If you do a little algebra, saying that m∠*A* + m∠*B* = m∠*B* + m∠*C*, and subtracting m∠*B* from both sides, you can show that ∠*A* and ∠*C* are the same size. In any parallelogram, opposite angles are congruent.

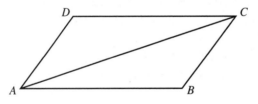

Draw a diagonal in any parallelogram and you form two triangles. Let's draw diagonal $\overline{AC}$ in parallelogram *ABCD*. That will form △*ACD* and △*CAB*. Because $\overline{DC}$ ∥ $\overline{AB}$, alternate interior angles ∠*DCA* and ∠*BAC* are congruent. Because $\overline{AD}$ ∥ $\overline{BC}$, ∠*DAC* and ∠*BCA* are congruent. That gives you two pairs of congruent angles, and $\overline{AC}$ is between those angles in both triangles and equal to itself. If you rotate △*CAB* so that ∠*BAC* sits on top of ∠*DCA* and ∠*BCA* sits on top of ∠*DAC*, not only will the shared side match itself, but you'll see $\overline{AB}$ matching $\overline{CD}$ and $\overline{BC}$ matching $\overline{AD}$. That tells you that the opposite sides are not only parallel, but also congruent.

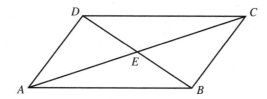

Drawing one diagonal in a parallelogram divides it into two matching triangles. When both diagonals are drawn in the parallelogram, it makes four triangles. If we call the point where the diagonals intersect point $E$, we can show that $\triangle ADE$ matches $\triangle CBE$. $AD = BC$, because the opposite sides of the parallelogram are congruent. m$\angle ADE$ = m$\angle CBE$ and m$\angle DAE$ = m$\angle BCE$ because the opposite sides are parallel, so alternate interior angles are congruent. That tells you how to match up the parts of the triangles, and you'll see the other parts match up as well. If the triangles are the same size and shape, $DE = EB$ and $AE = EC$, so the diagonals of the parallelogram bisect each other.

The family of parallelograms is made up of many different types of parallelograms. Some have only the properties of parallelograms we've covered so far, but others are special in one or more ways.

**CHECK POINT**

For each quadrilateral described, decide if there is enough information to conclude that the quadrilateral is a parallelogram.

1. In quadrilateral $ABCD$, $\overline{AB} \parallel \overline{CD}$ and $\overline{BC} \parallel \overline{AD}$.

2. In quadrilateral $PQRS$, with diagonal $\overline{PR}$, $\angle QRP \cong \angle SPR$ and $\angle QPR \cong \angle SRP$.

3. In quadrilateral $FORK$, $\angle F \cong \angle K$ and $FO = RK$.

4. In quadrilateral $LAMP$, with diagonals $\overline{LM}$ and $\overline{AP}$ intersecting at $S$, $\triangle ALS \cong \triangle PMS$ and $\triangle AMS \cong \triangle PLS$.

5. In quadrilateral $ETRA$, with diagonals $\overline{ER}$ and $\overline{TA}$ intersecting at $X$, $TX = RX$ and $EX = AX$.

# Rectangles

A rectangle is a parallelogram in which adjacent sides are perpendicular, and so it has four right angles. Because the rectangle is a parallelogram, it has all the properties of a parallelogram, but also the special property that all four of the angles are the same size. It's an equiangular parallelogram.

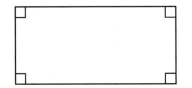

*A **rectangle** is a parallelogram with four right angles.*

You've probably worked with rectangles before, because there are so many rectangles all around us. You're reading this on a rectangular page, and you could be sitting in a room built from rectangles.

Every rectangle is a parallelogram, so its opposite sides are parallel and congruent, its consecutive angles are supplementary ($90° + 90° = 180°$), and its opposite angles are congruent. Those are properties of every parallelogram. The rectangle also has consecutive congruent angles, because all angles are congruent.

In any parallelogram, a diagonal makes two congruent triangles. That's still true in a rectangle, but those congruent triangles will both be right triangles. In any parallelogram, the diagonals bisect each other, and because the rectangle is a parallelogram, that's still true. But the rectangle has another special property. The diagonals of a rectangle are congruent. In most parallelograms, which sort of slant to one side, there's a long diagonal and a short one, but in a rectangle, both diagonals are the same length.

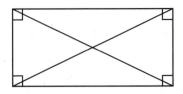

Suppose that in rectangle *ABCD*, the diagonals intersect at *E*. If *BE* = 8, how long is *AE*? Because the diagonals are congruent and bisect each other, *AE* = *EC* = *BE* = *ED*. So if *BE* = 8, *AE* = 8 as well.

# Rhombuses and Squares

A *rhombus* is a parallelogram with four sides of the same length, an equilateral parallelogram. Because the rhombus is a parallelogram, it has all the properties of a parallelogram, and it still may have that lean to one side, if its angles are different sizes. It's an equilateral parallelogram, but not always an equiangular one.

A *square* is a parallelogram that is both a rhombus and a rectangle. Squares have four right angles and four equal sides. They are equilateral and equiangular.

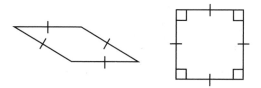

*A **rhombus** is a parallelogram in which all sides are congruent. A **square** is a parallelogram with four congruent sides and four right angles.*

In a rhombus or a square, drawing one diagonal will make two congruent triangles, just as it does in any parallelogram. In a rhombus, those triangles will be isosceles, and in a square, they will be isosceles right triangles. You know that in 45-45-90 right triangles the length of the hypotenuse is the length of a leg times the square root of two. That means that the length of the diagonal of a square will be the length of a side times radical two. Because the square is a rectangle, both diagonals are the same length.

When you draw both diagonals in a rhombus or a square, a bunch of things happen. The diagonals bisect each other, just as they do in any parallelogram. Because the sides of the rhombus (or square) are all the same length, you can figure out that the diagonals of any rhombus, including a square, are perpendicular to one another.

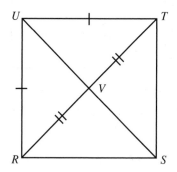

In *RSTU*, *RU = UT*, *RV = VT*, and *UV* equals itself, so $\triangle RVU$ and $\triangle TVU$ match each other. That means $\angle RYU$ is congruent to $\angle TVU$, but they're also supplementary, so they must both be 90°. If the diagonals cross at right angles, they're perpendicular.

For each quadrilateral described, choose the best label from parallelogram, rectangle, rhombus, or square.

6. In quadrilateral *FORT*, $\overline{FO} \perp \overline{OR}$, $\overline{OR} \perp \overline{RT}$ and $\overline{OR} \parallel \overline{FT}$.

7. In quadrilateral *CAMP*, $CA = AM = MP = CP$ and $\overline{AM} \perp \overline{MP}$.

8. In quadrilateral *VASE*, diagonals $\overline{VS}$ and $\overline{AE}$ are congruent, but sides $\overline{VA}$ and $\overline{AS}$ are not.

9. In quadrilateral *SOAP*, $\overline{SO} \parallel \overline{AP}$ and $\overline{AO} \parallel \overline{SP}$.

10. In quadrilateral *COLD*, diagonals $\overline{CL}$ and $\overline{OD}$ are perpendicular bisectors of one another, but they are not congruent.

# Trapezoids

A *trapezoid* is a quadrilateral with one pair of parallel sides and one nonparallel pair. It looks like a triangle that lost its head. The parallel sides are called the bases, and the nonparallel sides are called the legs.

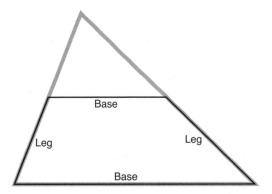

You might be able to see, with what you know about parallel lines and transversals, that the angles formed by the bases and one of the nonparallel sides are supplementary. These are called consecutive angles, but it's important to remember that the term "consecutive angles" applies only to the two angles at either end of a leg and not to the two angles at either end of a base. Those are base angles. The consecutive angles of a trapezoid are supplementary, but the base angles are not.

If the nonparallel sides are congruent, the trapezoid is an *isosceles trapezoid*. You might have guessed that you get an isosceles trapezoid by cutting the top off an isosceles triangle, so you won't be surprised to hear that the base angles of an isosceles trapezoid are congruent, just as the base angles of an isosceles triangle are congruent. But you might be surprised to know that the trapezoid has two sets of base angles: the ones at each end of the longer base and the ones at each end of the shorter base. The two angles in each set are congruent, and the angles in one set are supplementary to the angles in the other set.

*A **trapezoid** is a quadrilateral in which one pair of opposite sides are parallel. A trapezoid is an* ***isosceles trapezoid*** *if the nonparallel sides are the same length.*

The line segment joining the midpoints of the nonparallel sides is called the *median* of the trapezoid. The median is parallel to the bases. Its length is the average of the bases.

**DEFINITION**

The **median** of a trapezoid is a line segment that connects the midpoints of the two nonparallel sides.

Like all quadrilaterals, a trapezoid has two diagonals. In an isosceles trapezoid, the diagonals are congruent. Diagonals of other trapezoids are not congruent. Diagonals of a trapezoid do not bisect one another. That belongs only to parallelograms.

**CHECK POINT**

11. In trapezoid $ABCD$, $\overline{AC} \parallel \overline{BD}$ and $\overline{MN}$ is a median. $AC = 14$ cm and $BD = 30$ cm. How long is median $\overline{MN}$?

12. In trapezoid $FIVE$, $\overline{IV} \parallel \overline{FE}$ and m$\angle F = 59°$. What is the measure of $\angle I$?

13. In trapezoid $TEAR$, $\overline{EA} \parallel \overline{TR}$ and $TE = AR$. If $\angle E = 107°$, find the measures of $\angle A$ and $\angle R$.

14. In trapezoid $ZOID$, $\overline{ZD} \parallel \overline{OI}$. m$\angle Z = 83°$ and m$\angle I = 97°$. If $ZO = 4$ cm, how long is $\overline{ID}$?

15. In trapezoid $PQRT$, $\overline{PT} \parallel \overline{QR}$ and $\overline{MN}$ is a median. If $MN = 17$ inches and $PT = 21$ inches, how long is $\overline{QR}$?

# Perimeter of Quadrilaterals

It's easy to calculate the perimeter of quadrilaterals. Just as with triangles, you simply add up the lengths of the sides. Most figures don't even have a special formula for perimeter.

For a rectangle, you see the formula $P = 2L + 2W$. Because the opposite sides are the same length, you have two of the lengths and two of the widths. For a square or a rhombus, that becomes $P = 4s$. All the sides are the same length, and you have four of them.

Let's look at a story problem using perimeter.

Suppose Marianna wants to build a fence around her vegetable garden. If the garden is a rectangle 30 feet long and 15 feet wide, and fencing costs $1.25 per foot, how much will it cost to fence the garden?

The perimeter of a rectangle is $P = 2L + 2W$, so she will need $(2 \times 30) + (2 \times 15) = 60 + 30$, or 90 feet of fencing. 90 feet of fencing at $1.25 per foot will cost $90 \times 1.25 = \$112.50$. Marianna's fence will cost $112.50.

# Area of Quadrilaterals

The area of a quadrilateral is a little trickier to calculate than the perimeter, but once you know a few formulas, it's not hard at all. The area of any parallelogram is found by multiplying the base times the height ($A = bh$). In a triangle, the height must be measured as the perpendicular distance between a vertex and the base. In a parallelogram, the height is the perpendicular distance between two bases.

 **MATH TRAP**

Don't confuse the side with the height. In most parallelograms, the sides are not perpendicular, so you can't use a side as a height. The exception, of course, is rectangles, where the sides do meet at right angles.

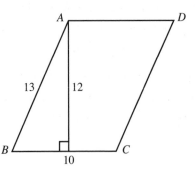

To find the area of parallelogram *ABCD*, it's not enough to know that *AB* = 13 meters and *BC* = 10 meters. You could use either of those measurements as the base, but neither of them is a height. You need to know the perpendicular distance between two bases. If you know that the height, drawn to *BC*, is 12 meters, then you can find that the area is 12 × 10 or 120 square meters.

> **WORLDLY WISDOM**
>
> Remember that area is always measured in square units. That's because you multiply a length in units (such as inches, feet, or meters) by another length in the same units. Units multiplied by units equals units squared, or square units.

That area formula of base times height ($A = bh$) will work for any parallelogram, including rectangles, rhombuses, and squares. But there are other area formulas, specific to those special parallelograms, which may be useful from time to time. Some are just another way of saying things. Because a rectangle has adjacent sides that are perpendicular, the sides, usually called the length and the width, are the base and the height, so you see the formula for the area of a rectangle as $A = L \times W$. The formula for the area of a square is special, because it is a rectangle, so you could use $A = L \times W$, but all four of its sides are the same length, so the formula becomes $A = s^2$.

> **WORLDLY WISDOM**
>
> The habit of referring to the exponent 2 as "squared" comes from the fact that the area of a square is the length of a side to the second power.

Let's look at a story problem using area.

Suppose you want to buy carpet for a room that measures 15 feet by 20 feet. The room is a rectangle, so you multiply 15 by 20 and go to the carpet store, knowing you need 300 square feet of carpet. When you arrive, you find that carpet is sold by the square yard, not the square foot. What now?

You have two choices. Your first option is to start over and figure out the measurements of your room in yards instead of feet. You know there are 3 feet in a yard, so 15 ÷ 3 = 5 yards and 20 ÷ 3 = $6\frac{2}{3}$ yards. Then you can multiply 5 by $6\frac{2}{3}$ to find out you need $33\frac{1}{3}$ square yards.

Your second option is to take your 300 square feet and divide by the number of square feet in a square yard. A square yard is a yard times a yard, or 3 feet times 3 feet, so it's 9 square feet. Your 300 square feet divided by 9 is $33\frac{1}{3}$ square yards. In both cases, you arrived at the same answer.

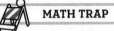

**MATH TRAP**

To get units squared, you need to multiply units by units, that is, feet times feet or meters times meters. But if you don't pay attention to whether the units match, you could be multiplying feet times inches, and that doesn't give you square feet or square inches. It just gives you the wrong answer. Make sure your units match.

The rhombus is usually thought of as the most unusual parallelogram, and its area formula is the most different from the others. It's built from the fact that the diagonals divide the rhombus into four congruent right triangles. The area of each right triangle is one-half of the product of its base and height, and it's a right triangle, so the legs are the base and height. Each leg is half of a diagonal, so if we call the the diagonals $d_1$ and $d_2$, the area of each right triangle is $\frac{1}{2} \cdot \frac{1}{2}d_1 \cdot \frac{1}{2}d_2 = \frac{1}{8}d_1d_2$. Since there are four triangles that make up the rhombus, the area of the rhombus is $4 \times \frac{1}{8}d_1 \cdot d_2 = \frac{1}{2}d_1 \cdot d_2$, or one-half the product of the diagonals.

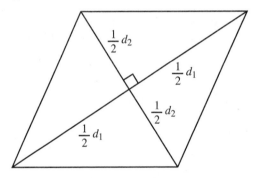

How would you use that formula? Suppose you need to find the area of a rhombus with diagonals 12 cm and 20 cm. You could go through a lot of calculating to try to find the length of the side, but you'd still need the height. Instead, remember that the area of a rhombus is half the product of the diagonals. The area of the rhombus is $\frac{1}{2} \cdot 12 \cdot 20 = 120$ cm$^2$.

So you can use the formula $A = bh$ for all parallelograms if you know the base and the height. That covers the $A = LW$ for the rectangle and the $A = s^2$ for the square. You can use the fancy formula for the rhombus if you know the lengths of the diagonals. But what about a trapezoid? It's not a parallelogram, so your parallelogram formula doesn't work, but you might need to find its area anyway. How?

Well, it just so happens that if you draw the median of a trapezoid and cut along the median, you can flip the top piece over and set it next to the bottom piece. When you do that, the two pieces will fit together to make a parallelogram.

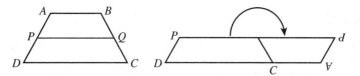

The height of the parallelogram is half the height of the trapezoid. The base of the parallelogram is the sum of the length of the long and the short base of the trapezoid. So the area of a trapezoid is equal to half the height multiplied by the sum of the bases.

$$A = \frac{1}{2}h\left(b_1 + b_2\right)$$

**WORDLY WISDOM**

Some people remember this formula as the average of the bases times the height, or the length of the median times the height.

Try using that formula to figure out the height of a trapezoid with an area of 40 square centimeters and bases 3 cm and 5 cm. Plug those numbers into the formula.

$$A = \frac{1}{2}h\left(b_1 + b_2\right)$$

$$40 = \frac{1}{2}h\left(3+5\right)$$

$$40 = \frac{1}{2}h \cdot 8$$

$$40 = 4h$$

$$h = 10$$

The height of the trapezoid is 10 centimeters.

**CHECK POINT**

16. Find the perimeter and area of a square with a side of 17 cm.

17. Find the perimeter and area of a rectangle 18 inches long and 9 inches wide.

18. Find the area and perimeter of parallelogram $ABCD$ if $AB = CD = 7$ inches, $AC = BD = 21$ inches, and the height from $B$ perpendicular to $\overline{AC}$ is 3 inches.

19. Find the perimeter and area of a rhombus with sides 5 inches long if the diagonals measure 6 inches and 8 inches.

20. The area of a parallelogram with a height of 48 cm is 3,600 square centimeters, and its perimeter is 250 cm. Find the lengths of the sides of the parallelogram.

# Polygons with More than Four Sides

Although triangles and quadrilaterals are the polygons you meet most often and the ones about which you have the most information, there are others. The other polygons have more sides, more vertices, more angles, and more diagonals, and there are a few rules they all follow. Even those rules will depend on the number of sides, but once you know that, you can figure out some things.

Remember that the name of a polygon is determined by how many sides it has. When there isn't a particular name for a polygon with that number of sides, you just tack *–gon* on to the number of sides. A polygon with seventeen sides would be a seventeen-gon.

### Naming Polygons by Sides

| Sides | 5 | 6 | 7 | 8 | 9 | 10 |
|-------|---|---|---|---|---|----|
| Name | Pentagon | Hexagon | Heptagon | Octagon | Nonagon | Decagon |

## Number of Diagonals

The more sides a polygon has, the more vertices it has, so the more diagonals you can draw. A triangle with three vertices has no diagonals, but a quadrilateral with four vertices has two diagonals. How many are there in a pentagon? Or a hexagon?

Let's start with a pentagon, with five sides and five vertices. You can start a diagonal from any one of the five vertices, but once you pick the starting point, you have four vertices left. Of those four, you can only draw diagonals to two others. If you try to draw to the other two, you'll be tracing over a side. So you have five places to start, but you can't end where you started, and you can't trace over a side, so you have only two places to end. You can draw two diagonals from that vertex, and there are five vertices, so there are 10 diagonals in a pentagon, right?

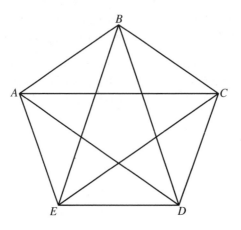

Not right. If your vertices were *A*, *B*, *C*, *D*, and *E*, the diagonal from *C* to *A* is the same diagonal you drew from *A* to *C*. In fact, there are only five diagonals in a pentagon because of that duplication.

The hexagon has six starting places and three ending places, but again you have to divide by two because of the duplicates. Six times three is 18, divided by two is nine diagonals in a hexagon. If a polygon has *n* sides, it will have $\dfrac{n(n-3)}{2}$ diagonals.

## Sum of the Angles

The sum of the measures of the three angles of a triangle is 180°. A quadrilateral can be split into two triangles by drawing one diagonal, and you can see that the total of the measurements of the four angles in a quadrilateral is 360°. What about polygons with more sides?

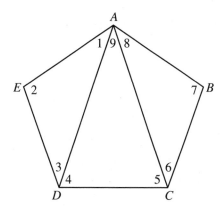

If you draw all the diagonals from one vertex, you break the polygon up into triangles. In this pentagon, drawing all the possible diagonals from one vertex divides the pentagon into three triangles. The angles of the pentagon are split up, but adding up the angles of all the triangles will make up the angles of the pentagon. The triangles each have a total of 180°, so the total for the three triangles is 540°. The total of the five interior angles of the pentagon is 540°.

The total number of degrees in the interior angles of a polygon with *n* sides is 180° times the number of triangles you create by drawing the diagonals from one vertex. The number of triangles is two less than the number of sides, so the total number of degrees is 180(*n* − 2).

21. Find the number of diagonals in an octagon (8 sides).

22. Find the total of the measures of all the interior angles in a nonagon (9 sides).

23. If a hexagon is regular, find the measure of any one of its interior angles.

24. If the interior angles of a polygon add up to 900°, how many sides does the polygon have?

25. If a polygon has a total of 119 possible diagonals, how many sides does it have?

## Area of Regular Polygons

When it comes to the area of polygons with more than four sides, there aren't a lot of rules you can follow. In most cases, you have to find a way to break the polygon up into smaller polygons, maybe triangles, find the area of those and add them all up. It's something, but most times you won't have the information you need to find all those areas.

There is one situation where you can follow a rule, and that's when you're looking for the area of a *regular polygon*. A regular polygon is one that is equilateral and equiangular. It has sides that are all the same length and angles that are all the same size.

**DEFINITION**

A polygon is **regular** if all sides are the same length and all angles are congruent.

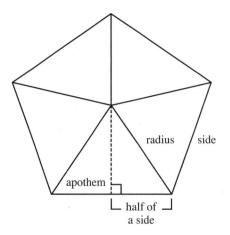

A regular polygon has a center point, and the radius, or distance from the center to a vertex, is the same no matter what vertex you go to. If you draw all the radii, you break the polygon into congruent triangles, one for each side of the polygon. If you can find the area of one of those triangles, you can multiply by the number of sides to find the area of the polygon.

Each of the triangles is isosceles, because all the radii are equal. The altitude from the center vertex to a side of the polygon is called the *apothem*, and it's a super segment that is a median and angle bisector as well as an altitude. That means that if you know the length of a side and the radius, you can find the length of the apothem. Or if you know a side and the apothem, you can find the radius. Or if you know the radius and the apothem, you can find the side. How? The Pythagorean theorem, $a^2 + b^2 = c^2$. When you draw the radii and an apothem, you create a right triangle whose legs are the apothem and half a side and whose diagonal is the radius. If you know any two of those, you can find the third.

> 📖 **DEFINITION**
>
> The **apothem** of a regular polygon is a line segment from the center of the polygon perpendicular to a side.

Suppose you have a regular pentagon with an apothem equal to 4 cm and a radius of 5 cm. You could use the Pythagorean theorem to find out that half the side is 3 cm, so the whole side is 6 cm. Once you have the measures of the apothem and a side, you can find the area of one triangle. $A = (\frac{1}{2} \times 6) \times 4 = 12$ square centimeters. There are five of those triangles, so the area of the pentagon is $5 \times 12 = 60$ square centimeters.

The area of a regular polygon with $n$ sides of length $s$ and apothem $a$ is given by the formula $A = n\left(\frac{1}{2}as\right)$, but since $n\left(\frac{1}{2}as\right) = \frac{1}{2}a \cdot ns$ and $ns$ is the perimeter of the regular polygon, the formula is often given as $A = \frac{1}{2}aP$ where $P$ is the perimeter.

**CHECK POINT**

26. Find the area of a regular pentagon with sides 8 cm long and an apothem 5 cm long.

27. Find the area of an octagon with a perimeter of 40 inches and an apothem of 5 inches.

28. Find the area of a regular decagon if each of the 10 sides measures 2 meters and the apothem is 1.5 meters.

29. The perimeter of a regular hexagon is 42 inches and its area is 84 square inches. How long is its apothem?

30. A regular pentagon has an area of 1,080 square inches and an apothem of 18 inches. How long is each side?

## The Least You Need to Know

- Parallelograms have both pairs of opposite sides parallel. Area formula: $A = bh$.
- Rhombuses are parallelograms with four equal sides. Area formula: $A$ = the product of the diagonals divided by 2.
- Rectangles are parallelograms with four right angles. Area: $A = lw$. Squares are rectangles that are also rhombuses. Area formula: $A = s^2$.
- Trapezoids are quadrilaterals with one pair of parallel sides. Isosceles trapezoids have nonparallel sides congruent and base angles congruent. Area formula: $A = \frac{1}{2}h$ multiplied by the sum of the bases.
- Regular polygons have congruent sides and congruent angles.
- Total of the interior angles of a polygon with $n$ sides is $180(n - 2)$.

# Circles

Triangles, quadrilaterals, pentagons, hexagons, and all the figures you've learned about so far have one thing in common: line segments. All polygons are built from line segments that form their sides and meet to form their angles. But the world of geometry isn't all straight lines.

Around the world of geometry, you'll find lots of things that are round, things that curve. The basic shape that forms the basis of all things that curve is the circle. In this chapter, you'll meet the circle and all the lines and angles that are associated with it. You'll get to know what those lines tell you about the circle and see how they relate to the familiar triangle. You'll learn how to measure the angles, and find out how they relate to the circle and to one another. And just as you learned to find the perimeter and area of a polygon, you'll explore how to measure the distance around a circle and how to measure the space inside it.

## In This Chapter

- Exploring the properties of circles
- Measuring angles in and around the circle
- Investigating lines and segments that interact with circles
- Finding the area and circumference of circles
- Connecting a circle drawn on the coordinate plane to an equation

# The Language of Circles

Take a piece of string and tie a pencil to one end. Pin the other end of the string to the center of a sheet of paper, grab the pencil and pull it out as far as the string will go. With the pencil point on the paper, move the pencil wherever you can, keeping the string taut. The pencil moves around the point where you pinned the string, tracing a curve that comes around to meet itself. That curve is a circle.

A *circle* is defined as the set of all points at a fixed distance from a given point. The point is called the *center* of the circle. That's the spot where you pinned down the string. The fixed distance that determines the size of the circle is called the *radius*. That's the length of the string. (The plural of radius is radii.) All radii of a circle are the same length, because every point on the circle is the same distance from the center.

 **DEFINITION**

> The collection of all points that sit at a certain distance, called the **radius**, from a set point, called the **center**, forms the shape called a **circle**.

If you mark two or more points on a circle, you divide the circle into sections. Each portion or part of the circle is called an *arc*. An arc that is less than half the circle is a *minor arc*. An arc that is greater than half the circle is a *major arc*. A half circle is called a *semicircle*.

 **DEFINITION**

> An **arc** is a portion of a circle. A **semicircle** is half a circle, a **minor arc** is smaller than a semicircle, and a **major arc** is larger than a semicircle.

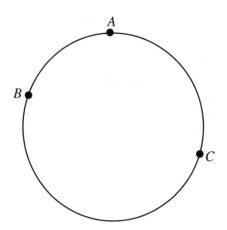

Arcs are named by the points at each end of the arc, with an arc symbol over the top, but that isn't always enough information. The notation $\overset{\frown}{AC}$ means the minor arc that is the shorter path along the circle from $A$ to $C$, but you might want to name the major arc that goes around the circle in the other direction from $A$ to $C$. To show that you want a major arc, include another letter of a point on the major arc to show which way you want to travel. So $\overset{\frown}{ABC}$ names the major arc from $A$ through $B$ to $C$.

Most of the units of measurement you use were designed for lines, so when you get to circles, you need a different way of measuring. The type of measuring you want to do in circles is much more like measuring an angle than measuring a line, so you borrow the system of degrees for use in circles. Degrees measure the rotation, how far you turn around a point. A circle turns all the way around its center point until it meets itself again. That full rotation is 360°. There are 360° in a full circle and 180° in a semicircle. Every arc can be assigned a measurement as a number of degrees, according to what part of the circle it is. Minor arcs are less than 180°, and major arcs are more than 180°. A quarter of the circle is one-fourth of 360°, or 90°.

One circle, all by itself, is pretty, but there's not a lot to say about it. Its center controls where it is, its radius controls how big it is, and it's always the same shape. If you have two or more circles, they can begin to interact a little bit. You can have two circles that just barely touch each other and have just one point in common. Those are *tangent* circles.

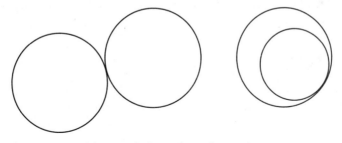

*Tangent circles* touch at only one point.

One circle can be tangent to another from outside or from inside. If the circles are externally tangent, they can be the same size or different sizes, but to be internally tangent, they'll have to be different sizes or they'll just sit right on top of one another. If a smaller circle is placed inside a larger one so that their centers are aligned, they won't touch each other, but they are *concentric* circles, circles with the same center.

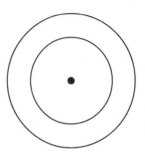

*Concentric circles* are circles with the same center.

### MATH IN THE PAST

Not every curve is a circle or part of a circle. Ancient mathematicians looked at the different shapes that are seen when a cone is cut. If the cut goes straight across, you see a circle, but if it slants a little, or a lot, you see different shapes. One of them is an oval shape that mathematicians call an ellipse. It's sort of a squashed circle.

### CHECK POINT

Complete each sentence.

1. An arc less than half a circle is a _____ arc.

2. The distance from the center point to any point on the circle is called the _____.

3. Two circles with the same center are _____ circles.

4. If two circles touch each other at just one point, the circles are _____.

5. An arc that is exactly half the circle is called a _____.

# Segments and Angles

When an angle is drawn so that its vertex is at the center of a circle, it's called a central angle. A *central angle* is an angle formed by two radii. The measure of a central angle is equal to the measure of the arc between the endpoints of its sides, its intercepted arc. The measure of a central angle is equal to the measure of its intercepted arc, and the measure of the arc is the measure of the central angle that cuts it off. The two measurements are interchangeable.

### DEFINITION

A **central angle** is an angle with its vertex at the center of a circle whose sides are radii.

A *chord* is a line segment whose endpoints lie on the circle. The longer the chord, the closer it will be to the center of the circle. Short segments will fit near the edge, but longer ones will have to go closer to the center. The *diameter* is the longest chord of a circle. It passes through the center of the circle. The diameter is twice as long as the radius.

When you draw a chord, you divide the circle into two portions, two arcs, usually a major arc and a minor arc. You think of the minor arc as being the arc that belongs to that chord. If you draw two chords in the same circle and the two chords are the same length, then the arcs they cut off will be the same size. Congruent chords cut off congruent arcs.

> 📖 **DEFINITION**
>
> A **chord** is a line segment that connects two points on a circle. The **diameter** is the longest chord that can be drawn in a circle and passes through the center.

If you draw two radii, they both come from the center point, and they form a central angle. If you draw two chords that come from the same point on the circle, you form an angle that also intercepts an arc, but you can see that it's smaller than the central angle that intercepts that same arc. An *inscribed angle* is an angle whose sides are chords and whose vertex lies on the circle. In the figure, central angle $\angle AOB$ and inscribed angle $\angle ACB$ both intercept arc $\overset{\frown}{AB}$, but you can see that $\angle ACB$ is smaller.

> 📖 **DEFINITION**
>
> An **inscribed angle** is an angle with its vertex on the circle whose sides are chords.

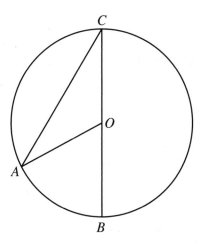

If you look at $\triangle AOC$, you can see it's an isosceles triangle because radii $AO = OC$, and that means that $m\angle OAC = m\angle OCA$. In addition, $\angle AOB$ is an exterior angle of the triangle, so $m\angle AOB = m\angle OAC + M\angle OCA = 2m\angle OCA$. That means the central angle is twice the size of the inscribed angle, or, in other words, the inscribed angle is half the size of the central angle. The measure of an inscribed angle is equal to one-half the measure of its intercepted arc.

Suppose that in circle $O$, $\overline{OA}$ and $\overline{OB}$ are radii and $\overline{AC}$ and $\overline{BC}$ are chords. If $\overset{\frown}{AB} = 50°$, find m$\angle AOB$ and m$\angle ACB$.

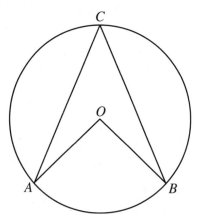

$\overset{\frown}{AB}$ is the intercepted arc for both angles. m$\angle AOB$ is a central angle, so its measure is the same as the measure of the arc. m$\angle AOB = 50°$. m$\angle ACB$ is an inscribed angle, so its measure is half the measure of the arc m$\angle ACB = 25°$.

> 💡 **WORLDLY WISDOM**
>
> An angle inscribed in a semicircle is a right angle.

When two chords intersect within a circle, they form four angles, which are labeled with numbers in the figure. Vertical angles are congruent, so $\angle 1 \cong \angle 3$ and $\angle 2 \cong \angle 4$. You might look at the picture and think that arc $AC$ is smaller than arc $BD$ (and you'd be right), and so you might wonder how the two angles could have the same measurement.

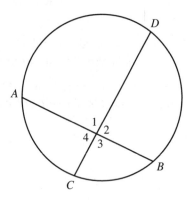

Draw chord $AD$ to make a triangle. $\angle 4$ is an exterior angle of that triangle, and so it's equal to m$\angle DAB$ + m$\angle CDA$. Those are inscribed angles, so m$\angle 4$ = m$\angle DAB$ + m$\angle CDA$ = $\frac{1}{2}AC + \frac{1}{2}DB$.

But then you notice that $\angle 2$ is also an exterior angle of that triangle and equal to exactly the same thing. The measure of an angle formed by two chords (and the measure of its vertical angle partner) is one-half the sum of the two intercepted arcs. To find the measure of an angle formed by two chords, average the arcs intercepted by the two vertical angles.

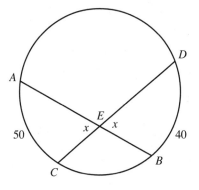

If two chords intersect in the circle as shown, and you want to find the value of $x$, you just need to realize that the two vertical angles whose measure is $x$ intercept arcs of 40° and 50°. Then $x = \dfrac{1}{2}(40 + 50) = 45°$.

---

**CHECK POINT**

6. $\angle TOP$ is a central angle in circle $O$, and arc $\overparen{TP}$ measures 48°. Find the measure of $\angle TOP$.

7. Points $M$, $A$, and $N$ lie on circle $O$. Arc $\overparen{MN}$ measures 78°. Find the measure of $\angle MAN$.

8. Diameter $\overline{SI}$ meets chord $\overline{IT}$ at point $I$ on circle $P$. Arc $\overparen{TS}$ measures 36°. Find the measure of $\angle SIT$.

9. In circle $P$ described in question 8, chords $\overline{IT}$ and $\overline{ST}$ meet each other at point $T$ and meet the ends of diameter $\overline{SI}$. Find the measure of $\angle ITS$.

10. Chords $\overline{LT}$ and $\overline{GP}$ intersect at point $A$ inside circle $O$. Arc $\overparen{LG}$ measures 56° and arc $\overparen{PT}$ measures 82°. Find the measure of $\angle GAL$ and the measure of $\angle PAL$.

# Lines and Angles

Radii, diameters, and chords are line segments. They have endpoints that are either on the circle or at the center of the circle. Lines that go on forever can also create angles that interact with the circle.

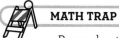

**MATH TRAP**

Remember that the center is not a point of the circle. The circle is points that are a set distance from the center.

A *secant* is a line that contains a chord. It is a line that intersects the circle in two distinct points. A *tangent* is a line that touches the circle in exactly one point. Just the way tangent circles only touched each other at one point, a tangent line only touches the circle at one point.

Secants and tangents also make angles on or around the circle. Two tangents, each of which touches the circle in just one point, can cross outside the circle, making an angle. It will look like an ice cream cone (the two tangents) holding a scoop of ice cream (the circle.) Two secants can cross outside the circle to form the vertex of the angle, and then each intersects the circle twice, dividing it into four arcs. Or a secant and a tangent can make an angle with its vertex outside the circle, with the secant cutting the circle twice and the tangent just touching the side.

**DEFINITION**

A **secant** is a line that intersects the circle it two different points. A **tangent** is a line that only touches the circle at one point.

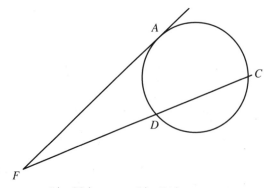

*Line FC is a secant. Line FA is a tangent.*

When these lines start making angles, there are a lot of angles and a lot of arcs that you can look at, but the good news is that all three kinds of angles come down to the same measuring rule. Let's look at them one by one, starting with the angle formed by two secants.

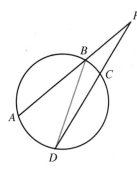

The figure shows two secants that cross at *P*. One cuts the circle at *B* and then again at *A*. The other crosses the circle at *C* and at *D*. You want the measurement of ∠*P*. To create some angles you do know how to measure, draw *BD*. That will make Δ*BDP* and exterior angle ∠*ABD*. m∠*ABD* = m∠*P* + m∠*BDC*, so m∠*P* = m∠*ABD* − m∠*BDC*. Both ∠*ABD* and ∠*BDC* are inscribed angles, so each is equal to half its intercepted arc. That means that the measure of ∠*P* is half of $\overset{\frown}{AD}$ minus half of $\overset{\frown}{BC}$, or half of the difference between $\overset{\frown}{AD}$ and $\overset{\frown}{BC}$. The measure of an angle formed by two secants is half the difference of the two arcs it intercepts.

Angles formed by two secants, a tangent and a secant, or two tangents will intercept two arcs. The arc nearest to the vertex of the angle is the smaller of the two. The two secants, as you can see in the figure, cut off a small arc $\overset{\frown}{BC}$ the first time they cross the circle and a larger one $\overset{\frown}{AD}$ the second time they cross the circle. The measure of the angle is one-half the difference of the two arcs it intercepts.

Two tangents are drawn to the circle from point *P*, as shown in the figure. Each tangent touches the circle at one point, *A* or *B*. The circle is broken into two arcs: the minor arc $\overset{\frown}{AB}$ and the major arc $\overset{\frown}{ACB}$. The measure of ∠*P* will be half the difference between the major arc and the minor arc, $m\angle P = \frac{1}{2}\left(m\overset{\frown}{ACB} - m\overset{\frown}{AB}\right)$.

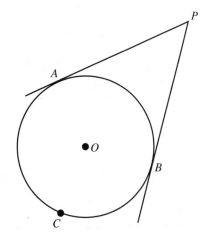

When an angle is formed by two tangents, the circle is broken into two arcs. When two secants make the angle, there are four arcs, but only the two that fall between the secants are used in the measurement. When a secant and a tangent team up to make an angle, you get three arcs, $\overset{\frown}{AC}$, $\overset{\frown}{CD}$, and $\overset{\frown}{AD}$, as shown below. The two you want are the two between the tangent and the secant, $\overset{\frown}{AC}$ and $\overset{\frown}{AD}$. The measure of $\angle F$ is $\frac{1}{2}\left(\overset{\frown}{AC}-\overset{\frown}{AD}\right)$.

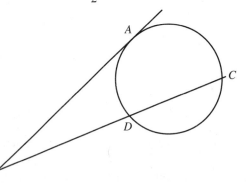

**WORLDLY WISDOM**

All the rules about the measurement of angles in and around circles can be boiled down to four rules, according to where the vertex of the angle is. If the vertex is at the center, the angle equals its intercepted arc. If the vertex is on the circle, the angle is half the intercepted arc. If the vertex is inside the circle, the angle is half the sum of the arcs, and if the vertex is outside the circle, the angle is half the difference of the arcs.

One last type of angle that you should know about is the angle that's formed by a line and a segment, or specifically a tangent and a radius or diameter, or a tangent and a chord.

A tangent just touches the circle at one point. When you draw a chord from that point of tangency $T$ to some other point on the circle, like $C$, you divide the circle into two arcs, minor arc $\overset{\frown}{TC}$ and major arc $\overset{\frown}{TDC}$, and you create two angles, $\angle ATC$ and $\angle BTC$. The measure of each angle is half its intercepted arc. $m\angle ATC=\frac{1}{2}\overset{\frown}{TDC}$ and $m\angle BTC=\frac{1}{2}\overset{\frown}{TDC}$. The two arcs add up to the whole circle, or 360°, so half of each one will add to 180°, which sounds just right because the angles are a linear pair and linear pairs are supplementary.

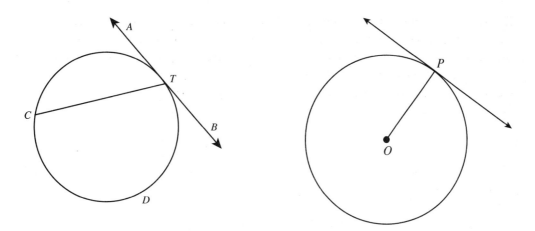

If the chord you drew happened to be a diameter, both intercepted arcs would be 180°, so both angles would be 90°. That means that a diameter drawn to the point of tangency is perpendicular to the tangent line, and because a radius is half the diameter, a radius drawn to the point of tangency is perpendicular to the tangent.

**CHECK POINT**

11. Two tangents $\overrightarrow{PA}$ and $\overrightarrow{PE}$ meet at point $P$ outside circle $O$ and touch the circle at $A$ and $E$. Minor arc $\overset{\frown}{AE}$ measures 140° and major arc $\overset{\frown}{ATE}$ measures 220°. Find the measure of $\angle APE$.

12. Two lines pass through point $A$, outside circle $O$. One intersects circle $O$ at points $E$ and $F$. The other line intersects the circle at $S$ and then $T$. The measure of $\overset{\frown}{ES}$ is 12°, and the measure of $\overset{\frown}{FT}$ is 52°. Find the measure of $\angle FAT$.

13. Tangent $\overrightarrow{RT}$ touches circle $O$ at point $A$ and meets chord $\overline{AC}$. Minor arc $\overset{\frown}{AC}$ measures 72°. If m$\angle RAC$ > m$\angle TAC$, find the measure of $\angle RAC$.

14. Tangent $\overrightarrow{PQ}$ and secant $\overrightarrow{PRS}$ form an angle of 15°. If arc $\overset{\frown}{QS}$ measures 160°, find the measure of arc $\overset{\frown}{QR}$.

15. Tangents $\overline{PA}$ and $\overline{PB}$ touch circle $O$ at points $A$ and $B$, dividing the circle into two arcs. The measure of the larger arc is twice that of the smaller arc. Find the measure of $\angle P$.

# Area and Circumference

When you explored polygons, whether triangles, quadrilaterals, or polygons with more sides, you always wanted to know two things: how to quickly find the distance around the edges of the polygon (its perimeter), and how to measure the area enclosed by the polygon. You want to be able to do these same things for the circle.

The *circumference* of a circle is the distance around the circle. The circumference is the circle's equivalent of the perimeter of a polygon. The distance around a circle is a little more than three times the diameter of the circle. The formula for the circumference of a circle is $C = \pi d$, where $d$ is the diameter of the circle and $\pi$ (*pi*) is a constant approximately equal to 3.14159.

> **DEFINITION**
>
> The **circumference** of a circle is the distance around the circle. The word *circumference* comes from Latin. *Circum* is the preposition for around, and *ferre* is a verb that means carry. If you carry something around the circle, you trace out the circumference.

You sometimes see the formula as $C = 2\pi r$, because $r$ is the radius of the circle, and the diameter is twice the radius. Because $\pi$ is an irrational number, many times you'll give your answer as a number times $\pi$, for example, $9\pi$. If you use an approximate value, you'll want to say "approximately equal" when you give your answer. For most questions, you can use 3.14 or $\frac{22}{7}$ as approximate values of $\pi$.

Suppose the circumference of circle $O$ is $40\pi$ cm. To find its radius, use the circumference formula $C = \pi d = 40\pi$. The diameter must be 40 cm, and therefore the radius is half of that, or 20 cm.

The *area* of a circle, the space enclosed by the circle, can be approximated in a number of ways. One is to sandwich it between a polygon whose sides are tangent to the circle, called a *circumscribed* polygon, and a polygon whose vertices are on the circle, called an *inscribed* polygon. The area of the circle will be larger than the area of the inscribed polygon but smaller than the area of the circumscribed polygon. To get a good approximation, however, you would have to use polygons with many sides and that makes for difficult calculations.

> **DEFINITION**
>
> A polygon is **circumscribed** about a circle if the polygon surrounds the circle with each side tangent to the circle. A polygon is **inscribed** in a circle if each of its vertices lies on the circle.

Here's a way to approximate the area of the circle that's a little easier. Cut the circle into wedges, like a pie. Line up the wedges, alternating point up and point down. It should look like this:

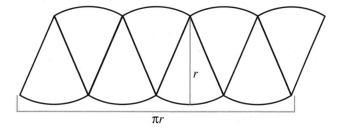

The rearranged area looks a lot like a parallelogram. The height of the parallelogram is the radius of the circle, and the base is half the circumference, or $\pi r$. The area is base times height, or $r$ times $\pi r$, which equals $\pi r^2$. The area of a circle is the product of $\pi$ and the square of the radius.

To find the area of a circle whose radius is 7 inches, you could just square 7 to get 49 and say the area is $49\pi$ square inches. That's absolutely true, but if you don't really have a sense of what $49\pi$ means, you could use $\pi \approx \dfrac{22}{7}$. $A = \pi r^2 \approx \dfrac{22}{7} \cdot 7^2 \approx 154$ square inches.

---

**CHECK POINT**

16. Find the area of a circle with a radius of 9 cm.

17. Find the circumference of a circle with a diameter of 12 inches.

18. Find the area of a circle with a diameter of 32 cm.

19. Find the circumference of a circle with an area of $36\pi$ square meters.

20. Find the area of a circle with a circumference of $24\pi$ feet.

# Circles in the Coordinate Plane

When you looked at the coordinate plane, you were interested mostly in the graphs of lines and the connection between the line and its equation. When you draw a circle on the coordinate plane, you can see that its equation has to be more complicated than the equation of a line, because the pattern of where the points fall is more complicated. But it turns out that you already know enough to figure out that equation, and it all goes back to the definition.

A circle is defined as the set of all points at a fixed distance from a given point. Let's keep things simple in the beginning and use the origin as the center point. That will mean that the circle is all the points that are *r* units from the point (0,0) for whatever radius *r* you pick. Let's have $r = 5$ for our example.

Finding the first few points is easy. From the origin, count 5 up and put a point at (0,5). Then count 5 down from the origin and put a point at (0,-5). Do a left and right move and put points at (5,0) and (-5,0). With those points you can imagine what the circle might look like, but it would be easier to draw it with a few more points.

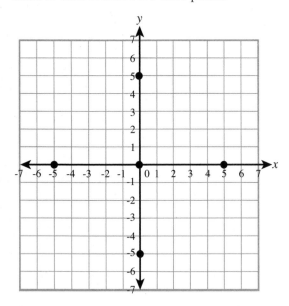

Do you remember the distance formula? It's the one that says the distance between two points $(x_1, y_1)$ and $(x_2, y_2)$ is $d = \sqrt{(x_2 - x_1)^2 + (y_2 - y_1)^2}$. And do you remember that the distance formula is just the Pythagorean Theorem in disguise?

You know that every point on the circle is 5 units away from the origin, so if you put a 3-4-5 right triangle with one vertex at the origin and one leg

along the *x*-axis, the other end of its hypotenuse will be 5 units from the origin.

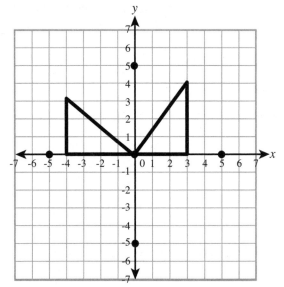

You can use that 3-4-5 right triangle like a measuring stick, putting it in different positions, and you get more points on your circle. When it looks like this, it will be easy to connect the dots.

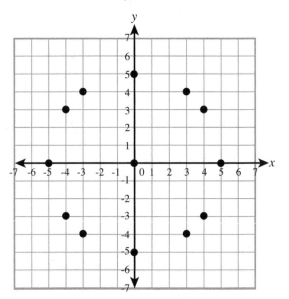

But let's go back to the <u>distance formula</u> for just another minute. $d = \sqrt{(x_2 - x_1)^2 + (y_2 - y_1)^2}$. You know one point is the origin, (0,0), and the distance is 5. Let's plug those in.

$$5 = \sqrt{(x_2 - 0)^2 + (y_2 - 0)^2}$$
$$5 = \sqrt{(x_2)^2 + (y_2)^2}$$

Now drop the little twos at the bottom of the $x$ and $y$, and then square both sides to get rid of the square root.

$$5 = \sqrt{x^2 + y^2}$$
$$5^2 = x^2 + y^2$$
$$25 = x^2 + y^2$$

Do you know what that is? It's the equation of the following circle—a circle with a center at (0,0) and a radius of 5.

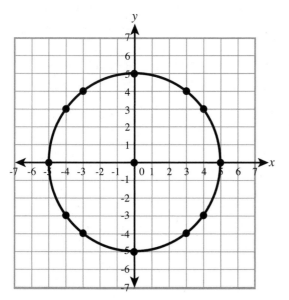

The equation of a circle with center (0,0) and radius $r$ is $x^2 + y^2 = r^2$.

If the center isn't at the origin, the equation gets a little more complicated, but in a way that makes sense if you remember that it came from the distance formula. Suppose you wanted a circle of radius 5 with its center at (1,4). The equation would be $(x - 1)^2 + (y - 4)^2 = 5^2$. The equation of a circle with center at $(h,k)$ and radius $r$ is $(x - h)^2 + (y - k)^2 = r^2$.

---

✏ ( **CHECK POINT** )

21. Write the equation of a circle with its center at the origin and a radius of 3 units.

22. Graph the circle represented by the equation $x^2 + y^2 = 64$.

23. Find the center and radius of the circle represented by the equation $(x - 8)^2 + (x - 3)^2 = 49$

24. Write the equation of a circle with center (4,9) and radius of 2 units.

25. Graph the circle represented by the equation $(x - 8)^2 + (y - 3)^2 = 36$.

## The Least You Need to Know

- A circle is the set of points at a fixed distance, called the radius, from a fixed point, called the center.
- A chord is a line segment that connects two points on the circle. A diameter is the longest chord in a circle and passes through the center.
- A secant is a line that contains a chord. A tangent touches a circle at just one point.
- A central angle has its vertex at the center of a circle and is equal to its intercepted arc. An inscribed angle has its vertex on the circle and is equal to half its intercepted arc.
- When two chords intersect inside a circle, each pair of vertical angles formed is equal to half the sum of the two intercepted arcs.
- Angles formed by two secants, two tangents, or a tangent and a secant are equal to half the difference of their intercepted arcs.
- The equation of a circle with center $(h,k)$ and radius $r$ is $(x - h)^2 + (y - k)^2 = r^2$.

# Surface Area and Volume

Most of the geometry we learn is plane geometry, which is geometry in two dimensions. But you don't live in a two-dimensional world. For life in three dimensions, you need to look at figures in space, not just on a flat surface or plane. Those 3-D figures are commonly referred to as solids (even if they're hollow).

The figures referred to as solids break down into several categories. There are figures with circular bases, figures with bases that are polygons, and figures that don't have a base. Those first two categories each divide into the objects that come to a point and those that don't. You'll get all those shapes organized and assign names to each. And you'll learn how to find the two most common measurements for each of them: volume and surface area.

## In This Chapter

- How to find the surface area and volume of prisms and pyramids.
- How to find the surface area and volume of cylinders and cones.
- How to find the surface area and volume of spheres.

# Measuring Solids

Line segments have length, a number that tells you the distance from one endpoint to the other. Length is measured in linear units, like inches, feet, centimeters, and meters. Plane figures like polygons and circles have two important measurements. The distance around the figure, perimeter for polygons and circumference for circles, is a measure of length, so it too is measured in linear units.

When you start measuring area, the space contained within the figure, you're not just measuring length. You're using measurements of length and width, working in two dimensions rather than one, so the area is measured in square units. The "square" designation makes sense if you remember you multiplied inches times inches or meters times meters to find that area. It makes sense that it's measured in square inches or square meters.

And now you want to measure three-dimensional figures. What kind of units will you need? That depends on what you're measuring. The length of an edge or the diameter of a circle that is a base of a 3-D figure will still be measured in linear units. If you want to talk about the *surface area* of a solid, that, like any area, is going to be measured in square units. When you begin to talk about *volume*, the measure of the amount this 3-D object might hold or the measure of the space it takes up, you're talking about length and width and height. That demands a new unit, a 3-D unit, called a cubic unit.

## DEFINITION

The **surface area** of a solid is the total of the areas of all the faces.

The **volume** of a solid is the measure of the space contained by the solid.

The simplest image of what we measure when we talk about volume is to imagine you have blocks, each 1 unit wide, 1 unit long, and 1 unit high. You can put your blocks into a box, packing them in rows, row after row until the bottom is full. Then you start making another layer of blocks on top of that. And another, and another, until the box is full.

Each of your blocks is a cube 1 unit by 1 unit by 1 unit. It's one unit cubed, or 1 cubic unit. The number of blocks you packed into the box is the volume of the box. It's how many blocks the box will hold. It's the volume of the box, in cubic units.

The problem, of course, is that you can't always find volumes by packing cubes into boxes. It takes too long, you don't always have enough cubes, and the cubes don't always fit nicely into the container. So you need other strategies for finding volume. Let's look at the different kinds of solids and work out strategies for each one.

# Prisms

First up are *polyhedrons*, figures made from polygons that meet at their edges. The polygons are called *faces* of the polyhedron.

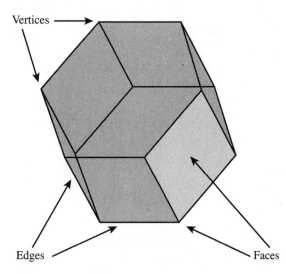

Vertices

Edges

Faces

A **polyhedron** is a solid constructed from polygons that meet at their edges. Each of the polygons is a **face** of the polyhedron.

Polyhedrons come in two varieties: prisms and pyramids. A *prism* is a figure that has a pair of parallel bases, connected to one another by parallelograms. There's enough math-speak in that sentence to scare even the bravest reader, so let's take it apart and look at some examples. Look at the book you're holding. It has rectangles for front and back covers, and those are parallel. The edges of the pages and the spine of the book form rectangles that connect the front cover to the back cover. When it's closed, the book is a *prism*, specifically a *rectangular prism*.

A *cube*, like a die, is an example of a prism, and a rather special one because all its faces are squares. Many boxes are good examples. You might have even seen fancy boxes with tops and bottoms that have six or eight sides. Those are prisms, too, but they're *hexagonal prisms* if the bases have six sides or *octagonal prisms* if the bases have eight sides. You may have seen a type of candy that's packed in a *triangular prism*. The ends of the box are triangles, and the three sides that connect the triangles are rectangles.

> **DEFINITION**
>
> A **prism** is a polyhedron with two parallel faces connected by parallelograms. Prisms take their names from the polygons that form the bases, such as a rectangle or octagon. The exception is the **cube**, which is a prism whose faces are all squares.

Most of the examples you see around you are what mathematicians call *right prisms*. That means that the two bases are positioned in line with one another so the sides meet the bases at a right angle. It is possible to make a prism that isn't right and seems to lean to one side, but you don't see them being used very often.

 **DEFINITION**

A prism is a **right prism** if the parallelograms meet the bases at right angles.

## Surface Area

Questions about the surface area can be answered by finding the area of each face of the solid and adding them up. It's often easier just to disassemble the prism, and then think about the area of the pieces. Do you have a triangular prism? That's two triangles for the bases and three rectangles around the sides. Check for the dimensions of each and go to work on the areas.

**MATH TRAP**

Don't automatically assume that all the parallelograms are the same size. If your bases are regular polygons, it might be true, but make sure you check first.

Suppose you're looking for the surface area of a triangular prism that stands 6 inches high, and the triangular bases are equilateral triangles with sides 3 inches long. You need to find the area of the two equilateral triangles and the area of the 3 rectangles.

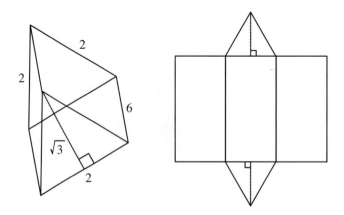

Each rectangle is 6 inches long and 2 inches wide, so each has an area of 12 square inches.

The equilateral triangles have a base of 2 inches, but you need a height, and for that you have to remember your 30-60-90 right triangles. The height of an equilateral triangle is half the length of a side times the square root of three, or in this case, $1\sqrt{3}$. The area of each triangle is $\frac{1}{2}(2)(\sqrt{3}) = \sqrt{3}$.

The total surface area is the two triangles plus the three rectangles or $2(\sqrt{3}) + 3(12) = 36 + 2\sqrt{3} \approx 39.46$ square inches. The area of the three rectangles is called the *lateral area*.

📖 **DEFINITION**

The **lateral area** of a prism is the total of the areas of the parallelograms surrounding the bases.

Coming up with a formula for that process is difficult because of all the different possibilities for the number of edges of the base and therefore the number of rectangles around it. Add the fact that the bases may or may not be regular polygons, and it's amazing that you can get any kind of formula, but you can.

Imagine a prism 6 inches high whose bases are 3-4-5 right triangles. The surface area is $2\left(\frac{1}{2} \cdot 3 \cdot 4\right) + 3 \cdot 6 + 4 \cdot 6 + 5 \cdot 6$. Let's rewrite those last three pieces, so the surface area looks like this: $2\left(\frac{1}{2} \cdot 3 \cdot 4\right) + 6(3 + 4 + 5)$. The sum in the parentheses, $3 + 4 + 5$, is the perimeter of the base. So you can say the surface area of the prism is two times the area of the base plus the height ($h$) times the perimeter of the base. If you use $B$ for the area of the base and $P$ for the perimeter of the base, the formula is $SA = 2B + hP$.

The surface area of a rectangular solid, is $SA = 2LW + 2Lh + 2Wh = 2LW + h(2L + 2W) = 2$(area of base) $+ h$(perimeter of base). Let's apply this to the rectangular prism shown.

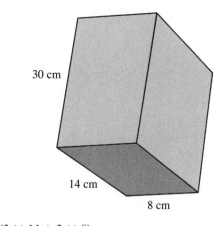

$$\begin{aligned} SA &= 2(14 \times 8) + 30(2 \times 14 + 2 \times 8) \\ &= 2(112) + 30(28 + 16) \\ &= 2(112) + 30(44) \\ &= 224 + 1320 \\ &= 1544 \end{aligned}$$

The surface area is 1,544 square centimeters.

---

> ✏️ **CHECK POINT**
>
> Find the surface area of each prism. $SA = 2B + hP$
>
> 1. A rectangular prism with edges of 15 cm, 24 cm, and 10 cm.
>
> 2. A triangular prism 8 inches high, with a base that is a right triangle with legs 5 inches and 12 inches long and a hypotenuse 13 inches long.
>
> 3. A hexagonal prism 42 cm high with a base that is a regular hexagon with a perimeter of 30 cm and an area of 65 square cm.
>
> 4. A pentagonal prism, 4 inches high, with a base that is a regular pentagon with sides 15 inches long and an area of 387 square inches.
>
> 5. A cube 17 cm on each edge.

## Volume

Rather than memorizing a lot of different volume formulas, remember that the volume of a prism is equal to the area of its base times its height. If you use $B$ for the area of the base, $V = Bh$.

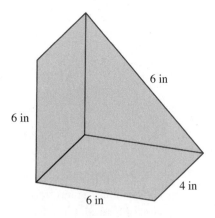

If you need to find the volume of a triangular prism 4 inches high, whose base is an equilateral triangle with sides 6 inches long, first you'll need to find the area of the base. Because it is an equilateral triangle, you can use the 30°-60°-90° triangle relationship to find the height. The altitude of the equilateral triangle is half the side times the square root of three, or $3\sqrt{3}$. The area of the triangle is $\frac{1}{2}bh = \frac{1}{2} \cdot 6 \cdot 3\sqrt{3} = 9\sqrt{3}$ square inches. Finally, the volume of the prism is the area of the base times the height, or $9\sqrt{3} \cdot 4 = 36\sqrt{3}$ cubic inches.

**WORLDLY WISDOM**

Volume is always measured in cubic units.

To find the volume of a rectangular prism with length of 15 inches, width of 9 inches, and height of 14 inches, first find the area of the base, $15 \times 9 = 135$ square inches, and multiply the area of the base by the height. $135 \times 14 = 1890$, so the volume of the rectangular prism is 1,890 cubic inches.

The volume of a rectangular prism is still $V = Bh$, but because the area of $B$ is length times width, the volume of a rectangular prism is sometimes written as $V = lwh$.

**CHECK POINT**

Find the volume of each prism. $V = Bh$

6. A cube 7 inches on each edge.

7. A rectangular prism with edges of 12 cm, 21 cm, and 15 cm.

8. A triangular prism 6 inches high, with a base that is a right triangle with legs 3 inches and 4 inches long and a hypotenuse 5 inches long.

9. A pentagonal prism, 8 inches high, with a base that is a regular pentagon with sides 15 inches long and an area of 387 square inches.

10. A hexagonal prism 50 cm high with a base that is a regular hexagon with a perimeter of 30 cm and an area of 65 square cm.

# Pyramids

A *pyramid* is a polyhedron formed by one polygon, called the base, surrounded by triangles that meet at a point. Like prisms, pyramids take their names from the polygon that forms the base. If the base is a pentagon, it's a pentagonal pyramid. If the base is a square, it's a square pyramid, and if the base is also a triangle, it's a triangular pyramid.

**DEFINITION**

A **pyramid** is a polyhedron composed of a polygon for a base surrounded by triangles that meet at a point.

## Surface Area

Like the prism, the pyramid can be broken down into its parts to find the surface area. If the base has $n$ sides, the surface area is the area of the base plus the area of the $n$ triangles that surround it. You can calculate each of the pieces separately and add them up.

> **MATH TRAP**
>
> You need to have the right slant on measurements in a pyramid. The word "height" gets used in a lot of different ways and doesn't always mean the same thing. In a prism, the height is one of the dimensions of the surrounding rectangles, because those rectangles stand straight up at right angles to the base. In a prism, the surrounding triangles are tilted. Their height, called the slant height, is longer than the height of the pyramid.

Usually when you're asked to find the surface area of a pyramid with a polygon as its base, that polygon is regular. All its sides are the same length and all its angles are equal. That makes finding its area a little easier. The area of a regular polygon is $\frac{1}{2}$ times the apothem times the perimeter.

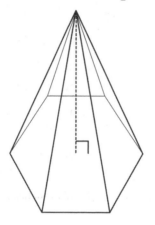

To find the surface area of a hexagonal pyramid, you have to know the area of the hexagon at the base and the area of the six surrounding triangles. If the hexagon is a regular hexagon, you can find the area if you know a side (or the perimeter) and the apothem. To find the area of the triangles, you need to know the side of the hexagon and the height of the triangle, which is the slant height of the pyramid. The lateral area is six times the area of one triangle, or $\frac{1}{2}$ times the perimeter times the slant height.

> **WORLDLY WISDOM**
>
> The area of a regular polygon with $n$ sides is the total of the $n$ triangles into which it is broken when you draw the radii from the center to each vertex. That translates to $n \times \frac{1}{2} \times$ apothem $\times$ a side. Because $n$ times a side is also the perimeter, the area of a regular polygon is $\frac{1}{2} \times$ apothem $\times$ perimeter.

It sounds like a lot of work, but here's how it breaks down. You need to know the perimeter ($p$) of the regular polygon at the base, the apothem ($a$) of that polygon, and the slant height ($l$) of the pyramid. Then you can find the surface area.

$$SA = (\frac{1}{2} \times a \times p) + (\frac{1}{2} \times p \times l)$$
$$= \frac{1}{2}p(a + l)$$

To find the surface area of a hexagonal pyramid with a perimeter of 18 inches, an apothem of 2 inches, and a slant height of 8 inches, plug those numbers into the formula.

$$SA = \frac{1}{2}p(a + l)$$
$$= \frac{1}{2}(18)(2 + 8)$$
$$= 90$$

The hexagonal pyramid has a surface area of 90 square inches.

**CHECK POINT**

Find the surface area of each pyramid. $SA = \frac{1}{2}p(a + l)$

11. A square pyramid 4 inches on each side with a slant height of 5 inches.

12. A triangular pyramid with a slant of 10 cm, whose base is an equilateral triangle 12 cm on a side, with an area of 62.4 square centimeters.

13. A pyramid with a slant height of 18 cm and regular pentagon as a base. The regular pentagon has a perimeter of 50 cm and an area of 172 square centimeters.

14. A hexagonal pyramid with a slant height of 10 inches. The regular hexagon that forms the base has a perimeter of 60 inches and an area of 260 square inches.

15. A square pyramid with a slant height of 13 inches and a side of 10 inches.

# Volume

The volume of a prism is $V = Bh$. The volume of a pyramid with the same base and the same height must be smaller, because those lateral faces went from rectangles to triangles and tipped inward. The volume of a pyramid is one-third of the area of the base times the height. If $B$ is the area of the base and $h$ is the height, the volume is $V = \frac{1}{3}Bh$.

Let's return to the hexagonal pyramid you looked at earlier, with $p = 18$ inches, $a = 2$ inches, and $l = 8$ inches. To find its volume, we need to know the area of the base and its height. (Remember, the height is different from the slant height.)

The area of its base is $B = \frac{1}{2}ap = \frac{1}{2}(2)(18) = 18$ square inches. For this pyramid, the height is about 7.75 inches. Plug those numbers into the formula.

$$V = \frac{1}{3}Bh = \frac{1}{3}(18)(7.75) \approx 46.5$$

The volume is about 46.5 cubic inches. Now consider a square pyramid with an edge 12 feet long and a height of 18 feet. If you plug those numbers into the formula, you get $V = \frac{1}{3}Bh = \frac{1}{3}(12^2)(18) = \frac{1}{3}(144)(18) = 864$. The pyramid will have a volume of 864 cubic feet.

### MATH IN THE PAST

The Great Pyramid of Giza, the largest of the three famous structures, is a square pyramid with a base 756 feet on each side and a height of 455 feet. That gives it a volume of over 86 million cubic feet. However, there isn't actually that much room inside it, because much of the space is occupied by the stones from which it was built.

### CHECK POINT

Find the volume of each pyramid. $V = \frac{1}{3}Bh$

16. A square pyramid with a slant height of 13 inches and a side of 10 inches.

17. A square pyramid 4 inches on each side with a slant height of 5 inches.

18. A triangular pyramid with a slant of 10 cm, whose base is an equilateral triangle 12 cm on a side, with an area of 62.4 square centimeters.

19. A pyramid with a slant height of 18 cm and regular pentagon as a base. The regular pentagon has a perimeter of 50 cm and an area of 172 square centimeters.

20. A hexagonal pyramid with a slant height of 10 inches. The regular hexagon that forms the base has a perimeter of 60 inches and an area of 260 square inches.

# Cylinders

Prisms and pyramids are solids formed from polygons, but there are also solids whose bases are circles. The good news is that they have a lot in common with prisms and pyramids.

A solid with two parallel bases that are circles (instead of polygons) is called a *cylinder* and has a lot in common with prisms.

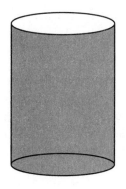

A **cylinder** is a solid with two circles as parallel bases and a rectangle wrapped around to join them.

## Surface Area

The surface area of a cylinder is the total of the areas of the two circles at the ends, plus the area of the rectangle that forms the sides. (Think about a label on a can. If you remove it from the can and unroll it, it's a rectangle.) The area of each circle is $\pi r^2$. The rectangle has a height equal to the height of the cylinder and a base equal to the circumference of the circle, so its area is circumference $\times$ $h$ or $2\pi rh$. The total surface area is $SA = \pi r^2 + 2\pi rh$.

Suppose that a can is going to be made in the shape of a cylinder with a radius of 4 inches and a height of 10 inches. How much metal will be needed to make the can? The metal needed to make the can is the surface area of the cylinder. Each of the two circular bases has an area of $16\pi$. The curved surface of the can unrolls to a rectangle whose base is the circumference of the circle, $8\pi$, and whose height is 10. The total surface area is $2 \times 16\pi + 8\pi \times 10 = 32\pi + 80\pi = 112\pi$ square inches. That's approximately 351.8 square inches.

## Volume

The similarities between prisms and cylinders hold up when volume is concerned. The volume of a prism is the area of the base times the height, and the volume of a cylinder is the area of its base, a circle, times its height. That makes the volume formula $V = \pi r^2 h$.

To find the volume of the cylinder whose diameter and height are both 4 inches, first find the radius. If the diameter = 4 inches, the radius = 2 inches, so the area of the base = $4\pi$ square inches. The volume is the area of the base times the height, or $4\pi \cdot 4 = 16\pi$ cubic inches. That's approximately 50.24 cubic inches.

---

Find the surface area and volume of each cylinder. $SA = \pi r^2 + 2\pi rh$ and $V = \pi r^2 h$

21. A cylinder 14 cm high with a radius of 5 cm.

22. A cylinder 8 inches high with a diameter of 6 inches.

23. A cylinder 2 m high with a circumference of $2\pi$ m.

24. A cylinder 82 cm high with a diameter of 90 cm.

25. A cylinder 20 inches high with a circumference of $20\pi$ inches.

# Cones

If cylinders behave a lot like prisms, it's not surprising that *cones* have a lot in common with pyramids. Like cylinders, cones have a circle as their base. Like pyramids, their lateral surface slopes to a point. Trying to unroll a cone is a bit trickier than taking the label off a can, as you did with the lateral area of a cylinder. If you try to unroll the cone's lateral area, you'll find it's part of a circle, but how big a part depends on the size of the base circle.

## Surface Area

The surface area of a pyramid is equal to the area of the base plus the combined area of all the surrounding triangles. That lateral area, you learned, was equal to $\frac{1}{2} \times$ perimeter $\times$ slant height. For a cone, the surface area is still the area of the base plus the lateral area. The base is a circle, so the area of the base is $\pi r^2$. If the slant height is $l$, the lateral area is a fraction of the area of a circle with area $\pi l^2$. What fraction? The fraction you create when you put the circumference of the base, $2\pi r$, over the circumference of a circle with radius equal to the slant height. $\frac{2\pi r}{2\pi l} \cdot \pi l^2 = \frac{r}{l} \cdot \pi l^2 = \pi rl$. The lateral area is $\pi rl$, so the surface area is $SA = \pi r^2 + \pi rl$.

To find the surface area of a cone with a radius of 8 cm and a slant height of 10 cm, find the area of the base. $\pi r^2 = \pi(8^2) = 64\pi$, and then calculate $\pi rl = \pi(8)(10) = 80\pi$. The surface area is $64\pi + 80\pi = 144\pi$ square centimeters.

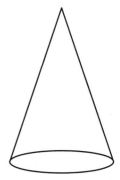

A **cone** is a solid with a circular base and a lateral surface that slopes to a point.

## Volume

The volume of a pyramid is one-third of the volume of a prism with the same base and height. Conveniently, the volume of a cone is one-third of the volume of a cylinder with the same base and height. For a cone, $V = \frac{1}{3}\pi r^2 h$.

To find the volume of a cone 9 meters high with a base that has a radius of 8 meters, follow the formula.

$$V = \frac{1}{3}\pi r^2 h$$

$$V = \frac{1}{3}\pi\left(8^2\right)(9) = \frac{1}{3}\pi(64)(9) = 192\pi$$

The volume is $192\pi$ cubic meters, or approximately 603.2 cubic meters.

> **CHECK POINT**
>
> Find the surface area and volume of each cone. $SA = \pi r^2 + \pi r l$ and $V = \frac{1}{3}\pi r^2 h$
>
> 26. A cone with radius 10 cm, height 24 cm, and slant height 26 cm.
>
> 27. A cone with diameter 8 inches, height 3 inches, and slant height 5 inches.
>
> 28. A cone with circumference $16\pi$ cm, height 6 cm, and slant height 10 cm.
>
> 29. A cone with radius 12 inches and height 5 inches. (Hint: Use the Pythagorean theorem to find the slant height.)
>
> 30. A cone with base area $324\pi$ cm and a height of 24 cm.

# Spheres

The last category of solids is the hardest to define and one you've probably known almost all your life. A sphere is not a polyhedron because it's not formed from polygons. It's not a cylinder because it doesn't have that lateral area. The sphere is the shape you'd produce if you could grab a circle by the ends of a diameter and spin it around that diameter so fast it blurred into something three dimensional. Put another way, it's a ball.

Officially, a sphere would be defined as the set of all points in space at a fixed distance from a center point. You can think of it as a 3-D circle, or a ball.

## Surface Area

The surface area of a sphere depends on the radius of the sphere. The formula $SA = 4\pi r^2$ gives you the surface area of a sphere of radius $r$.

An official soccer ball is a sphere with a circumference of 68 to 70 centimeters. That means its diameter is $\dfrac{68}{\pi}$ to $\dfrac{70}{\pi}$ centimeters, so its radius is $\dfrac{34}{\pi}$ to $\dfrac{35}{\pi}$ centimeters. That's between 10.8 and 11.1 centimeters. The surface area of the official ball will be at least $4\pi\left(\dfrac{34}{\pi}\right)^2 = 4\pi \cdot \dfrac{34^2}{\pi^2} = \dfrac{4 \cdot 34^2}{\pi} \approx 1471.9$ square centimeters and not more than
$4\pi\left(\dfrac{35}{\pi}\right)^2 = 4\pi \cdot \dfrac{35^2}{\pi^2} = \dfrac{4 \cdot 35^2}{\pi} \approx 1559.7$ square centimeters.

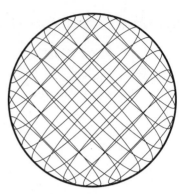

A **sphere** is the set of all points in space at a fixed distance from a center point.

## Volume

The volume of a sphere, like the surface area, depends on the radius of the sphere. The formula for the volume of a sphere is $V = \dfrac{4}{3}\pi r^3$. Volume is measured in cubic units, so it's not surprising that the radius is cubed. And given how round it is, the presence of $\pi$ isn't a surprise, either. Unfortunately, explaining where the $\dfrac{4}{3}$ came from requires more advanced math than I can include here, so you'll have to just believe me for that part.

The radius of the earth is about 3,959 miles. To find the volume of your home planet, calculate $V = \dfrac{4}{3}\pi r^3 = \dfrac{4}{3}\pi(3959)^3$.

$$V = \frac{4}{3}\pi r^3 = \frac{4}{3}\pi(3959)^3 \approx \frac{4}{3}\pi(62{,}052{,}103{,}080) \approx 82{,}736{,}137{,}440\pi$$

82 billion times pi is a very large number. It's approximately $2.6 \times 10^{11}$ or about 260 billion cubic miles.

**CHECK POINT**

31. Find the surface area of a sphere with a radius of 8 inches.

32. Find the volume of a sphere with a radius of 12 cm.

33. Find the surface area of a sphere with a diameter of 4 m.

34. Find the volume of a sphere with a diameter of 6 feet.

35. Find the radius of a sphere with a volume of $4500\pi$ cubic centimeters.

## The Least You Need to Know

- The surface area of a prism is the sum of the areas of the polygons that form the prism. The volume of a prism is the area of the base times the height.
- The surface area of a pyramid is the area of the base plus the areas of the triangles that surround it. The volume of a pyramid is one-third the area of the base times the height.
- The surface area of a cylinder is the area of the two circular bases plus the area of the rectangle that forms the lateral area, or $2\pi r^2 + 2\pi rh$. The volume of a cylinder is the area of the base times the height, or $\pi r^2 h$.
- The surface area of a cone is the area of the circular base plus the lateral area, or $\pi r2 + \pi rl$, where $l$ is the slant height. The volume of a cone is one-third $\pi r^2 h$.
- The surface area of a sphere is $4\pi r^2$. The volume of a sphere is $\frac{4}{3}\pi r^3$.

# Geometry at Work

## Areas of Irregular Figures

Once you know how to find the area of a certain type of polygon, it becomes a routine process. From time to time, you may have to work backward, if you're given the area and need to find one of the dimensions, but that's not all that difficult. The interesting problems are the ones that combine those basic formulas with a little bit of cut-and-paste thinking to figure out how to find the area of *that!*

It may be a figure that doesn't fit any rule when you take it as a whole, but can be broken down into sections that fit common rules nicely. For those problems, the trick is to break the figure up in a way that leaves you with figures whose areas you know how to find and whose dimensions you know or can find easily. Your strategy is to break the figure into parts, find their areas, and then put it all back together.

The other type of problem is one in which the overall outline of the shape does fit a rule but it has holes cut out of it. In those cases, the best tactic is to find the area of the whole shape and then subtract the area of the holes. "Whole minus holes" if you want a shortcut.

### In This Chapter

- How to use formulas and logic to find areas of irregular figures.
- How to use similar triangles to find measurements that can't be made directly.
- How to use trigonometric ratios to calculate measurements that can't be found by other means.

Let's take a look at one of these more interesting problems. The figure shown here is a quadrilateral, but it's definitely not part of the family of parallelograms. It's called a *concave polygon*. That dent from $X$ to $W$ to $Y$ is where it caves in. Concave polygons cave in. Officially, a polygon is concave if one of its diagonals falls outside the polygon. If you tried to draw a diagonal from $X$ to $Y$, it would be outside the quadrilateral.

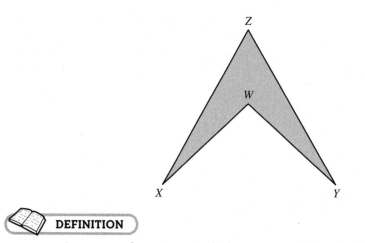

DEFINITION

A **concave polygon** is one in which one or more diagonals falls outside the polygon. A **convex polygon** is one in which all the diagonals fall inside the polygon.

All that is nice, but how do you find the area of something like that? You could divide and conquer. Draw $ZW$ and divide the figure into two triangles. Find the area of each and add them together.

Of course, to find the area of each triangle, you need to know the lengths of a base and the height to that base. If you knew that $ZW = 4$ cm and that an altitude from $X$ to the extension of $ZW$ measures 3 cm, you could calculate the area of $\triangle XWZ$ as $\frac{1}{2} \times 4 \times 3 = 6$ square centimeters. If the altitude from $Y$ is also 3 cm, you have the polygon broken into two triangles, and each triangle has an area of 6 square centimeters, so the quadrilateral has an area of 12 square centimeters.

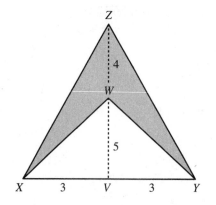

Or you could add more lines to the figure. Connect $X$ and $Y$, draw $\overline{ZW}$, and extend it to meet $\overline{XY}$ at $V$. Find the area of $\triangle XZY$ and the area of $\triangle XWY$ and subtract. If $XV$ and $YV$ are both 3 cm, $XY$ is 6 cm. $ZW$ is 4 cm, and if $WV$ is 5 cm, you can calculate the area of $\triangle XZY$ as $\frac{1}{2} \times 6 \times 9 = 27$ square centimeters, and the area of $\triangle XWY = \frac{1}{2} \times 6 \times 5 = 15$ square centimeters. Subtracting $27 - 15$ gives you the area of the quadrilaterals as 12 square centimeters.

It's not uncommon to be asked to find the area of part of a figure, the portion shaded in this picture. In problems like these, it can be confusing to find the measurements you actually need. In this target diagram, the two outer rings are marked as 1 inch wide, but there's no convenient formula for the area of a ring.

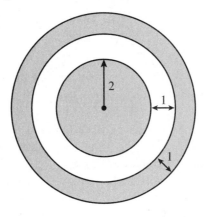

 **MATH TRAP**

When you're looking for the area of a shaded region, be careful to take the shape apart carefully. It's easy to include area you don't really want or miss area you do want if you rush to get an answer.

To find the shaded area, you're going to have to devise a way to get the area of that outer ring. You can deal with the circle in the middle later. It's a circle. You know what to do with that. Try thinking about the picture as a big gray circle with a smaller white circle sitting on top.

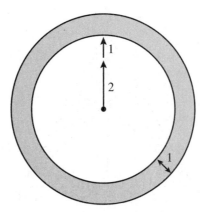

> **WORLDLY WISDOM**
>
> Taking the time to sketch out the portions of the figure before you begin calculating is time well spent. Especially if you're thinking of the figure as shaded and unshaded layers, it's wise to sketch the layers separately.

The smaller white circle has a radius of 3 inches. The gray circle underneath is an inch bigger, so its radius is 4 inches. The area of the big gray circle is $4^2$ times pi, or $16\pi$ square inches. The white circle has an area of $3^2$ times pi, or $9\pi$ square inches. The part of the gray circle peeking out from behind the white circle, the gray ring, has an area of $16\pi - 9\pi = 7\pi$ square inches.

Now that you know the area of that outer ring, you can bring back the gray circle in the middle. It has a radius of 2 inches, so its area is $4\pi$ square inches. Add that to the outer ring, and you have a shaded area of $7\pi + 4\pi = 11\pi$ square inches.

You've looked at circles in circles, what about squares in squares? In this figure, the area of square *ABCD* is 32 square centimeters. To find the area of the shaded region, work backward to find the dimensions you need. Square ABCD has an area of 32 square centimeters, so each of its sides is $\sqrt{32} = \sqrt{16}\sqrt{2} = 4\sqrt{2}$ centimeters. What you need, however, is to find the area of the gray isosceles triangles.

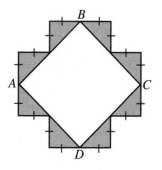

Each side of the square fits the hypotenuses of two triangles, so each hypotenuse is half of $4\sqrt{2}$, or $2\sqrt{2}$ centimeters. In an isosceles right triangle, with two legs the same length, the Pythagorean theorem says $a^2 + b^2 = a^2 + a^2 = 2a^2 = \left(2\sqrt{2}\right)^2$. That means that $2a^2 = 8$, $a^2 = 4$ and $a = 2$, so each leg of the little gray triangles is 2 centimeters long. The area of each little triangle is $\frac{1}{2} \times 2 \times 2$ or 2 square centimeters, and there are 8 triangles, making the total shaded area 16 square centimeters.

> **WORLDLY WISDOM**
>
> Leave any square roots in radical form at least until the end of your calculation. If you move to a decimal approximation sooner, you'll have to do your arithmetic with that decimal. That's more work, plus it's approximate to begin with, and you may need to round more at the end. Wait, keep the radical, and then, if you really need a decimal approximation, do it when all the calculation is done.

### CHECK POINT

Find the shaded area in each figure.

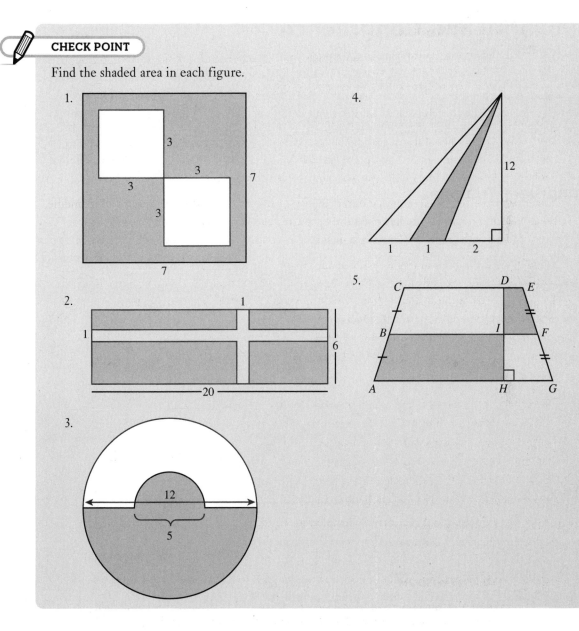

1.

3

3

3

3

3

7

7

2.

1

1

6

20

3.

12

5

4.

12

1    1    2

5.

C    D    E

B    I    F

A    H    G

# Similarity and Congruence

When we talk about how polygons relate to one another, we most often talk about congruence and similarity. Two polygons are *congruent* if they are the same shape and the same size. They're copies of one another, as if you ran one through a copier. Two polygons are *similar* if they're the same shape but different sizes. Think about enlarging a photo without distorting the image. Polygons of any number of sides can be congruent or similar, but most often you will encounter congruent triangles or similar triangles.

## Congruent Triangles

Triangles are *congruent* if they are the same shape and the same size. Because the size of the angles controls the shape of the triangle, in a pair of congruent triangles, you can match up angles that are the same size. In other words, corresponding angles are congruent. Because the length of sides controls size, corresponding sides are of equal length, or congruent.

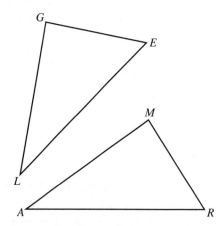

Two triangles are **congruent** if each pair of corresponding sides are congruent and each pair of corresponding angles are congruent.

If $\triangle ARM \cong \triangle LEG$, $\angle A$ and $\angle L$ will be the same size, $\angle R$ will have the same measurement as $\angle E$, and m$\angle M$ = m$\angle G$. Sides are congruent as well. $\overline{AR} \cong \overline{LE}$, $\overline{RM} \cong \overline{EG}$ and $\overline{AM} \cong \overline{LG}$.

The order in which the vertices of the triangle are named tells you what matches. You can sometimes tell what matches by looking at a picture, but unless you're told the measurements, or you can actually measure, you can't always be certain. The order in which the vertices of the triangles are named will tell the correct match-up.

Of course, that assumes that someone told you that the triangles were congruent. What if they asked you if the triangles were congruent? Technically, you'd have to measure all the sides and all the angles of both triangles and check to see that you can match up parts of the first triangle with parts of the second so that corresponding parts are congruent. That's what the definition of congruent triangles says.

That's a lot of measuring, and you might realize that, for example, you don't really have to measure the third angle of each triangle. If you measure the first two angles of both triangles and they match, the third angles will have to match as well, because the three angles of any triangle always add to 180°. That might get you to thinking about whether there's anything else you can skip.

It turns out that to be certain that two triangles are congruent, you only need to check three measurements from each triangle. You must always have at least one pair of congruent sides, to guarantee that the triangles are the same size. If you only have the measurements of one side of each triangle, and they are congruent, then you must have two pairs of matching angles to be certain the triangles are congruent. It can be any two pairs, because if you know that two pairs of angles are congruent, the third pair will match as well. If you measure two pairs of sides and find that they match, and you can show that the angles they form also match, you can be certain the triangles are congruent.

If you were to cut three sticks to set lengths and tried to put them together into triangles, you'd find out there's only one triangle you can make. If you measure all three pairs of sides in two triangles and they match, you don't need to measure any angles before you decide the triangles are congruent. Three pairs of congruent sides congruent is enough.

Each of these "minimum requirements" involves three pieces, and each one has a three-letter abbreviation.

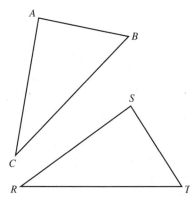

## Minimums Required To Be Certain Triangles Are Congruent

| Minimum Needed | Abbreviation | Example from $\triangle ABC \cong \triangle STR$ |
|---|---|---|
| Three pairs of corresponding sides congruent | SSS | $AB = ST$<br>$BC = RT$<br>$CA = RS$ |
| Two pairs of corresponding sides and the angle between them congruent | SAS | $AB = ST$<br>$BC = RT$<br>$\angle B = \angle T$ |
| Two pairs of corresponding angles and the side between them congruent | ASA | $\angle B = \angle T$<br>$\angle C = \angle R$<br>$BC = TR$ |
| Two pairs of corresponding angles and a pair of sides not between them congruent | AAS | $\angle B = \angle T$<br>$\angle C = \angle R$<br>$AB = ST$ |

Decide if the information given is enough to guarantee that the triangles are congruent. If so, state the rule that guarantees the congruence. If not, write "cannot be determined."

6. $\angle A \cong \angle X$, $\angle B \cong \angle Y$, $\overline{AB} \cong \overline{XY}$

7. $\overline{CT} \cong \overline{IN}$, $\angle A \cong \angle W$, $\angle C \cong \angle I$

8. $\overline{BI} \cong \overline{MA}$, $\overline{IG} \cong \overline{AN}$, $\angle B \cong \angle M$

9. $\overline{CA} \cong \overline{DO}$, $\overline{AT} \cong \overline{OG}$, $\overline{CT} \cong \overline{DG}$

10. $\overline{BO} \cong \overline{CA}$, $\angle O \cong \angle A$, $\overline{OX} \cong \overline{AR}$

## Similar Triangles

Triangles are similar if they are the same shape, but not necessarily the same size. Because they have to be the same shape, and angles control shape, corresponding angles are congruent. They don't have to be the same size, so you don't have to have corresponding sides congruent, but to change the size without distorting the triangle, corresponding sides will be in proportion. If the longest side of one triangle is twice the size of the longest side of the other triangle, then the shortest side of the first triangle will be twice the size of the shortest side of the second triangle. All the sides will be multiplied by the same number, called the scale factor.

Two triangles are **similar** if each pair of corresponding angles are congruent and corresponding sides are in proportion.

The **scale factor** of two similar triangles is the ratio of a pair of corresponding sides.

To conclude that triangles are similar, you must know that all three pairs of corresponding angles are congruent, and you must check that sides are in proportion. Just like with congruent triangles, however, there are shortcuts. The most common one for similar triangles is called AA. You must know that two angles of one triangle are congruent to the corresponding angles of the other. Of course, if two pairs of angles are congruent, the third pair will be congruent, too. The second shortcut is to check that two pairs of sides are in proportion and the angles between them are congruent to each other. The third is to check that all the sides are in proportion. In fact, AA is the one used most often.

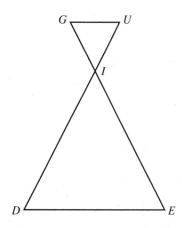

$\Delta GUI$ will be similar to $\Delta EDI$ if $\angle G \cong \angle E$ and $\angle U \cong \angle D$. The third angles are the pair of vertical angles, and vertical angles are congruent, so you know m$\angle GIU$ = m$\angle EID$. The corresponding sides will be in proportion, $\dfrac{GU}{DE} = \dfrac{UI}{DI} = \dfrac{IG}{IE}$. When you know that two triangles are similar, you can use a little squiggly line to show their relationship, like this: $\Delta GUI \sim \Delta EDI$.

---

✏️ **CHECK POINT**

Write the proportion for each pair of similar triangles.

11.  $\Delta ABC \sim \Delta XYZ$                14.  $\Delta MLN \sim \Delta LJK$

12.  $\Delta RST \sim \Delta FED$                15.  $\Delta ZXY \sim \Delta BCA$

13.  $\Delta PQR \sim \Delta VXW$

## Finding Missing Measurements

The primary reason that you want to know if a pair of triangles might be similar or congruent is that it will help you find a measurement you need but can't measure directly. If you know two triangles are congruent, you know you have two identical sets of measurements. Suppose $\Delta RST \cong \Delta MNP$. If $RS$ = 12 feet and $ST$ = 18 feet, you can find the length of $\overline{MN}$, even though it's in the other triangle. Because corresponding parts of congruent triangles are congruent, $MN$ = $RS$ = 12 feet.

If you know triangles are similar, you know angles are congruent and sides are in proportion. Suppose $\triangle RST \sim \triangle ABC$, $RS = 12$ feet, $BC = 6$ feet and $ST = 18$ feet, and you need to find the length of $\overline{AB}$. Use the fact that in similar triangles, corresponding sides are in proportion: $\dfrac{RS}{AB} = \dfrac{ST}{BC}$. Fill in the known lengths. $\dfrac{12}{x} = \dfrac{18}{6}$. Cross-multiply and solve. $18x = 72$, and $x = 4$. The length of $\overline{AB}$ is 4 feet.

# Indirect Measurement with Similar Triangles

Right triangles are often helpful in finding areas of irregular figures and in finding measurements indirectly, as you'll see a little later. Triangles of many different shapes can be used for indirect measurement, if you use what you know about similarity.

Remember the two triangle relationships, congruence and similarity? Either one can help you find a measurement you can't take directly, but in cases where you can't measure directly, you may not be able to create an exact copy to use congruent figures. Similarity is easier because you can make a smaller copy of the same shape and then use similar triangles to find the measurement of the larger version.

Similar polygons are the same shape but not necessarily the same size. Because their sides are in proportion, however, you can use a smaller version of the figure to calculate the measurements of a larger one.

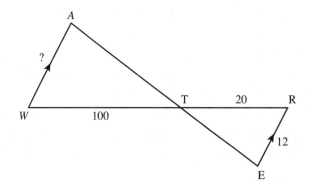

If you need to know how wide the river is from point $W$ on one bank to point $A$ on the other bank, you can't walk on water to take the measurement. But you can measure off 100 meters along one bank and mark point $T$, then go another 20 meters and mark point $R$. Create $\triangle RET$ to be similar to $\triangle WAT$ by making side $RE$ parallel to side $WA$, and $\angle WTA = \angle RTE$. (Okay, you probably don't walk around with a protractor in your pocket, but it's possible to do it.) Measure $RE$ and set up the proportion $\dfrac{WA}{RE} = \dfrac{WT}{TR}$. You can plug in what you know, and solve for $WA$.

$$\frac{WA}{RE} = \frac{WT}{TR}$$

$$\frac{x}{12} = \frac{100}{20}$$

$$20x = 1200$$

$$x = \frac{1200}{20} = 60$$

The river measures 60 meters from $W$ to $A$.

If you want to know if that big tree in your local park is as tall as the holiday tree in Rockefeller Center, you're not going to climb the tree with a yardstick. But you can go down to the park with a friend on a sunny day and use the sun, your friend, and similar triangles to help you measure the tree.

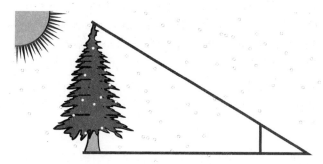

The sun will cause the tree to throw a shadow on the ground, and although you can't easily measure the tree, you can measure its shadow. Have your friend stand up nice and tall beside the tree and measure his height and the length of his shadow. The tree and its shadow form the legs of a right triangle (with a ray of sunlight as its hypotenuse). Your friend and his shadow make the legs of a similar right triangle. You can set up a proportion to find the height of the tree.

If your friend is 5 feet tall and casts a shadow 2 feet long, and the tree's shadow is 30 feet long, your proportion will look like this.

$$\frac{tree}{tree's\ shadow} = \frac{friend}{friend's\ shadow}$$

$$\frac{x}{30} = \frac{5}{2}$$

$$2x = 150$$

$$x = 75$$

Your tree is 75 feet tall, just a foot shorter than the 2013 Rockefeller Center tree.

CHECK POINT

Find the length of the specified segment.

16. $\triangle GHI \sim \triangle ARM$, $GH = 9$ ft, $GI = 8$ ft, $AR = 12$ ft. Find $AM$.

17. $\triangle JKL \sim \triangle DOG$, $JK = 17$ m, $JL = 25$ m, $DG = 30$ m. Find $DO$.

18. $\triangle ABC \sim \triangle XYZ$, $AB = 21$ cm, $BC = 54$ cm, $XY = 7$ cm. Find $YZ$.

19. $\triangle DEF \sim \triangle CAT$, $DE = 65$ in, $EF = 45$ in, $CA = 13$ in. Find $AT$.

20. $\triangle MNO \sim \triangle LEG$, $MN = x - 3$, $NO = 3$, $EG = 21$, $LE = 2x + 4$. Find $LE$.

# Indirect Measurement with Trigonometry

Trigonometry, or "triangle measurement," developed as a means to calculate lengths that can't be measured directly. It accomplishes this by using the relationships of sides of right triangles. These fundamental relationships of trigonometry are based on the proportions of similar triangles.

## Trigonometric Ratios

If you look at two right triangles, each with an acute angle of 25°, you can quickly prove that the two triangles are similar. The 25° angles are congruent and the right angles are congruent, so the triangles are similar. In fact, all right triangles containing an angle of 25° are similar, and you can think of them as a family. For any right triangle in this family, the ratio of the side opposite the 25° angle to the hypotenuse will always be the same.

That's not the ratio you usually think about with similar triangles, but it follows from the usual one. Most of the time, you'd say that the ratio of the side opposite the 25° angle in one triangle to the side opposite the 25° angle in the other is the same as the ratio of the hypotenuse to the hypotenuse. Let's write it this way: $\dfrac{\text{opposite1}}{\text{opposite2}} = \dfrac{\text{hypotenuse1}}{\text{hypotenuse2}}$. If you cross-multiply, divide both sides by hypotenuse1, and then divide both sides by hypotenuse2, here's what happens.

$$\frac{\text{opposite1}}{\text{opposite2}} = \frac{\text{hypotenuse1}}{\text{hypotenuse2}}$$

$$\text{opposite1 - hypotenuse2} = \text{opposite2 - hypotenuse1}$$

$$\frac{\text{opposite1 - hypotenuse2}}{\text{hypotenuse1}} = \frac{\text{opposite2 - \cancel{hypotenuse1}}}{\cancel{hypotenuse1}}$$

$$\frac{\text{opposite1 - hypotenuse2}}{\text{hypotenuse1}} = \frac{\text{opposite2}}{1}$$

$$\frac{\text{opposite1 - \cancel{hypotenuse2}}}{\text{hypotenuse1 - \cancel{hypotenuse2}}} = \frac{\text{opposite2}}{1 \text{ hypotenuse2}}$$

$$\frac{\text{opposite1}}{\text{hypotenuse1}} = \frac{\text{opposite2}}{\text{hypotenuse2}}$$

The ratio of the side opposite the 25° angle to the hypotenuse is the same in every right triangle in the 25° family. The ratios of other pairs of sides will remain constant throughout the 25° family, and it's not limited to just the 25° family. The ratios of pairs of sides are the same for all the right triangles in any family. Trigonometry takes advantage of that fact and gives a name to each of the possible ratios.

If the three sides of the right angle are labeled as the hypotenuse, the side opposite a particular acute angle, *A*, and the side adjacent to the acute angle *A*, six different ratios are possible. Three are commonly used. The other three are the reciprocals of the first three, and while it's nice to know they exist in case you ever need them, you can just deal with the three basic ratios called the *sine*, *cosine*, and *tangent*.

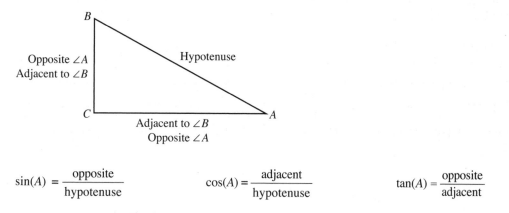

$$\sin(A) = \frac{\text{opposite}}{\text{hypotenuse}} \qquad \cos(A) = \frac{\text{adjacent}}{\text{hypotenuse}} \qquad \tan(A) = \frac{\text{opposite}}{\text{adjacent}}$$

Getting the definitions of the three principal trig ratios right is critical for your problem solving, so it helps to have a memory device, or mnemonic, to help you remember them. Many people like the word SOHCAHTOA to remember the trig ratios. It stands for **S**ine = **O**pposite/**H**ypotenuse; **C**osine = **A**djacent/**H**ypotenuse; **T**angent = **O**pposite/**A**djacent.

> **CHECK POINT**
>
> Triangle *ABC* has a right angle at $\angle B$. $AB = 12$ cm, $BC = 5$ cm and $AC = 13$ cm. Find each trig ratio.
>
> 21. sin $\angle A$
>
> 22. tan $\angle B$
>
> 23. cos $\angle A$
>
> 24. tan $\angle A$
>
> 25. sin $\angle B$

## Finding Missing Sides

With these three ratios and a table of their values (or a calculator that has keys for them), it is possible to solve for any unknown side of the right triangle if another side and an acute angle are known, or to find the angle if two sides are known.

> **MATH IN THE PAST**
>
> Once upon a time, students had to rely on tables to look up the values of the ratios for each family of right triangles, but now the sine, cosine, and tangent of an angle can be found with a few keystrokes on your calculator. Because the other three ratios don't have keys on the calculator, they're not used as often.

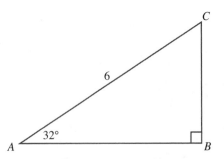

In right triangle $\triangle ABC$, hypotenuse $\overline{AC}$ is 6 cm long, and $\angle A$ measures 32°. To find the length of the shorter leg, it helps to first make a sketch to help you visualize the triangle. The shorter leg will be opposite the smaller angle, so if you draw the triangle and calculate the measure of each angle, you'll see that m$\angle A = 32°$, m$\angle B = 90°$, and m$\angle C = 58°$. You need to find the side opposite the 32° angle, or side $\overline{BC}$.

You need a ratio that talks about the opposite side and the hypotenuse (because that's the side you know). The sine is the one you want. $\sin(32°) = \dfrac{BC}{AC} = \dfrac{x}{6}$. From your calculator, you can find that $\sin(32°) \approx 0.5$, so $0.53 = \dfrac{x}{6}$ and $x \approx 3.2$. The side opposite the 32° angle is about 3.2 cm.

Remember that tree you measured by using similar triangles? You needed the help of a friend to make the second triangle, but what if your friend wasn't available? What would you have done?

With the help of trigonometry, you could find the height of the tree without your friend. Stand at the tip of the tree's shadow and measure the angle to the top of the tree. The height of the tree is the side opposite that angle, and the length of the shadow is the side adjacent to the shadow. The opposite side over the adjacent side forms the ratio called the tangent.

In this case, you'd probably find the angle to be about 68°, and you know the length of the tree's shadow is 30 feet. You can find tan(68°) on a table or calculator and set up the problem like this.

$$\tan(A) = \frac{\text{opposite}}{\text{adjacent}}$$

$$\tan(68.2°) = \frac{x}{30}$$

$$30\tan(68.2°) = x$$

$$30(2.5) = x$$

$$75 = x$$

Once again, the tree is 75 feet high.

**CHECK POINT**

26. $\triangle XYZ$ is a right triangle with right angle $\angle Y$ and hypotenuse $\overline{XZ}$ equal to 42 cm. If $\angle X$ measures 56°, find the length of side $\overline{XY}$ to the nearest centimeter.

27. $\triangle ABC$ is a right triangle with right angle $\angle C$, $\angle A = 46°$, and side $\overline{AC}$ is 42 cm long. Find the length of $\overline{BC}$ to the nearest centimeter.

28. In right $\triangle RST$, $\angle S$ is a right angle and $RT = 24$. If $\angle T$ measures 30°, find the length of $\overline{RS}$ to the nearest centimeter.

29. In right $\triangle ABC$ with right angle at $C$, $\angle B = 76°$, and side $\overline{BC}$ is 80 feet. Find the length of hypotenuse $\overline{AB}$.

30. In right $\triangle XYZ$ with right angle at $Y$, $\angle X = 32°$, and side $\overline{YZ}$ is 58 cm. Find the length of $\overline{XY}$ to the nearest centimeter.

## The Least You Need to Know

- To find the area of irregular figures, find the area of the whole figure and subtract the area of the unshaded portions, or divide the area into shapes whose areas can be calculated and add the results.

- When triangles are similar, corresponding sides are in proportion, and those proportions can be used to find the lengths of unknown sides.

- Trigonometric ratios, based on families of similar right triangles, can be used to find unknown sides of right angles.

- For any acute angle $A$ in a right triangle,
  $\sin(\angle A)$ = length of the opposite side divided by the length of the hypotenuse,
  $\cos(\angle A)$ = length of the adjacent side divided by the length of the hypotenuse, and
  $\tan(\angle A)$ = length of the opposite side divided by the length of the adjacent side.

# The State of the World

Your great adventure in the world of numbers is almost over, at least for now. Hopefully, you've enjoyed enough of this tour to want to visit at least some of these areas again.

What's left on the agenda? Well, when you've had a great adventure, you usually want to report on your experience when you get home. Telling people about what you've seen and what you've learned often means using some statistics, like the average rainfall in Thailand or the number of people who ride the Tube every day in London. Statistics are numbers that summarize a larger collection of information, and the heading "Statistics" also includes charts and graphs and other means of communicating the "big picture." This last part of your journey is about assessing risks and possibilities—probability—and summarizing what you've found out—statistics. These will help you make smart decisions and share what you've learned with the folks back home. Include a few photos—or charts and graphs—to keep things interesting for them.

# CHAPTER 18

# Probability

Math is all about the one right answer, right? Not so fast. It would be comfortable if everything had an absolutely certain right answer, but that's often not the case. One of the greatest skills math gives us is the ability to talk about possibility and uncertainty in an organized way. When you can't be certain that something is going to happen, being able to talk the chance of it happening and being able to assign a number to that chance is helpful.

In this chapter, you'll explore ways to assign a number to the likelihood that something will happen. You'll do that by comparing to all the things that could happen. In simple situations, that's all you need to do, but you'll see there are some rules to make the more complicated situations easier to handle. For all of this, you'll have to do a lot of counting, so let's start with a look at efficient methods of counting.

## In This Chapter

- The basics of counting
- Permutations and combinations
- How to find the probability of an event
- Probabilities when more than one event occurs

# Counting Methods

You know how to count, of course. If you had a bag of marbles, some red and some blue, and you wanted to know the chance that the first marble you pulled out (without looking) was a blue marble, the first thing you'd want to know is how many red marbles and how many blue marbles are in the bag. Unless it was a gigantic bag of marbles, you could answer that question by counting.

On the other hand, suppose you were going to buy a lottery ticket that asked you to pick five numbers from a group of 50, and you had to get all five correct to win. Your chance of winning will depend on how many ways there are to pick five numbers, and that counting problem is more complicated.

When you start to pick your number for the lottery ticket, you have 50 numbers to choose from. After you pick your first number, there are 49 left to choose from for the second number, then 48 for the third, 47 for the fourth, and 46 for the fifth. (That's assuming you have to pick five different numbers. If you're allowed to repeat, you have 50 choices each time.) And there's another question, too. Do you have to have the five numbers in just the right order? Is 12345 the same as 54321? Or are those different?

Big counting jobs raise many questions, but there are some rules that will help. Let's start with the most essential rule.

## Basic Counting Principle

The fundamental rule for quick counting is called the basic counting principle. It gives you a fast way of counting up the possibilities by multiplying the number of choices. Suppose you're getting dressed and you need to choose jeans or khakis and then pick a shirt from the six t-shirts in the drawer. You could wear any one of the six t-shirts with jeans, or any one of the six t-shirts with khakis. That's 12 possible outfits: 2 choices for pants times 6 choices for shirts.

If you have to do something that requires several choices, and you create a slot for each choice that needs to be made and fill each slot with the number of options for each choice, multiplying those numbers will tell you how many different possibilities you have.

If you have 5 shirts, 4 pairs of slacks, and 3 pairs of shoes, and you're comfortable mixing and matching any of them, how many different outfits can you make? Using the basic counting principle, create a slot for shirts, a slot for slacks, and a slot for shoes:

$$\underline{\hspace{1cm}} \quad \underline{\hspace{1cm}} \quad \underline{\hspace{1cm}}$$
shirts  slacks  shoes.

Then fill in the number of each that you have and multiply.

$$\underline{5} \times \underline{4} \times \underline{3} = 60$$
shirts  slacks  shoes    outfits

Now, if you're willing to do laundry often so everything is wearable, you could wear a different outfit every day for almost two months with just those few items. Add a couple of sweaters, and you'd be set for almost four months.

$$\underset{\text{sweaters}}{\underline{2}} \times \underset{\text{shirts}}{\underline{5}} \times \underset{\text{slacks}}{\underline{4}} \times \underset{\text{shoes}}{\underline{3}} = \underset{\text{outfits}}{120}$$

So what about that lottery ticket? Let's suppose you can't repeat numbers, but you do have to have them in the right order. The basic counting principle says you can multiply the number of options you have for each choice to find the total number of ways to pick five numbers from the pool of fifty.

$$\underline{50} \times \underline{49} \times \underline{48} \times \underline{47} \times \underline{46} = 254,251,200$$

That's over 250 million different ways to pick five numbers, and only one will win.

---

**CHECK POINT**

Your favorite salad bar has 2 kinds of greens, 5 different veggie toppings, 3 crunchy toppings, and 4 dressings. How many different salads can you make if you choose…

1. Greens and a dressing?

2. Greens, a veggie, and a dressing?

3. Greens, a veggie, a crunchy topping, and a dressing?

4. Greens, 2 different veggies, and a dressing?

5. Greens, 2 different veggies, 2 crunchy toppings, and a dressing?

---

## Permutations

A popular television game show has a game in which the contestant is told all the digits in the price of the very expensive prize (usually a car or a trip), but the contestant only wins the prize if he or she can put the digits in the correct order. Depending on how many digits there are, the chance of guessing correctly may be pretty good or pretty terrible.

If I told you that the digits in the price of your prize are 3 and 6, you're either going to guess 36 or 63. There are only two possible answers, and one of them is right. Basic counting principle says 2 choices for the first digit times 1 choice for the second digit gives you two possibilities. But you're not going to find a car for $36 or for $63. There will be quite a few digits in the price of the car or trip or other expensive prize. How many possibilities would there be if you had to put 5 digits in order? Or 6 digits?

This question of the number of ways you can arrange a group of things—numbers, letters, people, whatever—is the question of how many permutations there are. The word *permutation* means arrangement or ordering.

> **DEFINITION**
>
> A **permutation** is an arrangement or ordering of a group of objects.

You can answer the question about the number of permutations by using the basic counting principle. If you have to put five digits in order, there are $5 \times 4 \times 3 \times 2 \times 1 = 120$ permutations. This pattern of multiplication, from a number (in this case, 5) down to 1, is called a factorial. The symbol for factorial is an exclamation point. $5! = 5 \times 4 \times 3 \times 2 \times 1 = 120$.

The number of orders in which the 8 children on a soccer team could line up at the water fountain, or the number of different arrangements of 8 people taken 8 at a time, is $8! = 8 \times 7 \times 6 \times 5 \times 4 \times 3 \times 2 \times 1 = 40,320$. For any number $n$, the permutations of $n$ things taken $n$ at a time is $n!$

But what if you're not taking $n$ things at a time? For example, a baseball team needs to put 9 players on the field but might have 25 players on the roster. Now, if you know anything about baseball, you know that the players aren't all interchangeable, but still, how many permutations of 25 people taken 9 at a time are possible?

Your best strategy is to go back to the basic counting principle. Make nine slots and start filling them from 25 on down. $25 \times 24 \times 23 \times 22 \times 21 \times 20 \times 19 \times 18 \times 17$. Multiply that out and you get $\approx 7.4 \times 10^{11}$. That's approximately 740 billion possible lineups. Maybe it's a good thing the players aren't all interchangeable.

You're still talking about an arrangement, or permutation, but an arrangement of 9 out of the 25, not all 25. You don't want the entire 25!, but you do want the first 9 slots of it. You can write a formula for the permutations of $n$ things taken $t$ at a time as

$$_nP_t = \frac{n!}{(n-t)!}$$

In this case, the permutations of 25 things taken 9 at a time is

$$_{25}P_9 = \frac{25!}{(25-9)!}$$

$$= \frac{25!}{16!}$$

$$= \frac{25 \times 24 \times 23 \times 22 \times 21 \times 20 \times 19 \times 18 \times 17 \times \cancel{16 \times 15 \times 14 \times 13 \times 12 \times 11 \times 10 \times 9 \times 8 \times 7 \times 6 \times 5 \times 4 \times 3 \times 2 \times 1}}{\cancel{16 \times 15 \times 14 \times 13 \times 12 \times 11 \times 10 \times 9 \times 8 \times 7 \times 6 \times 5 \times 4 \times 3 \times 2 \times 1}}$$

Do you see how the factorial in the denominator cancels the part of the factorial in the numerator that you don't want? And isn't it good to know you don't have to write out all those factorials all the time?

So if you need the permutations of 7 things taken 4 at a time, the basic counting principle says you want $\underline{7} \times \underline{6} \times \underline{5} \times \underline{4}$, and the permutation formula says $_7P_4 = \dfrac{7!}{(7-4)!} = \dfrac{7!}{3!} = \dfrac{7 \times 6 \times 5 \times 4 \times \cancel{3 \times 2 \times 1}}{\cancel{3 \times 2 \times 1}}$. The approaches are different, but the result is the same.

**CHECK POINT**

6. What is 6!?

7. What is 7!?

8. What is 7! ÷ 6?

9. Find the permutations of 9 things taken 3 at a time.

10. Find the permutations of 10 things taken 4 at a time.

## Combinations

If 8 people are candidates in a school board election and the top two finishers will serve on the board, you might ask how many permutations of 8 people taken two at a time are possible. You can calculate that $_8P_2 = \dfrac{8!}{(8-2)!} = \dfrac{8!}{6!} = 8 \times 7 = 56$. There are 56 different orders of finish.

On the other hand, if both of the top two finishers will serve, regardless of which is first and which is second, the order in which they finish doesn't really matter. Mr. Smith and Ms. Jones is a different permutation from Ms. Jones and Mr. Smith, but they both give you the same school board members.

Because all the different arrangements of the same two people shouldn't count separately, the number of permutations is too big, but it can give you a place to start. There are 56 permutations, or arrangements, of 8 people taken 2 at a time, but fewer *combinations* or groups in which order doesn't matter. Specifically, there are half as many combinations as permutations.

**DEFINITION**

A **combination** is a way of selecting a number of objects from a larger group when the order of the objects is not significant.

What if the top three vote-getters were chosen for the school board, instead of the top two? There would be $8 \times 7 \times 6 = 336$ permutations of eight people taken three at a time, but how many combinations? If you were to list all the permutations, your list would include some permutations that include Mr. Smith, Ms. Jones, and Dr. Johnson. How many permutations on the list include these three people? That's the same as asking how many permutations are there of three people taken three at a time, and the answer is $3 \times 2 \times 1 = 6$. Each group of three people contributes 6 permutations to the total of 336. Because there are six arrangements of these same three people—six arrangements of any three people—the number of permutations is six times larger than the number of groups or combinations. Dividing the number of permutations by 6, or $\frac{_{12}P_3}{_3P_3}$, will give you the number of combinations. The combinations of $n$ things taken $t$ at a time is $_nC_t = \frac{_nP_t}{_tP_t} = \frac{n!}{(n-t)!} \div t! = \frac{n!}{t!(n-t)!}$.

The permutations of 10 things taken 5 at a time is

$$_{10}P_5 = \frac{10!}{(10-5)!} = \frac{10!}{5!} = \frac{10 \times 9 \times 8 \times 7 \times 6 \times \cancel{5 \times 4 \times 3 \times 2 \times 1}}{\cancel{5 \times 4 \times 3 \times 2 \times 1}} = 30{,}240.$$

The combinations of 10 things taken 5 at a time is

$$_{10}C_5 = \frac{_{10}P_5}{_5P_5} = \frac{10!}{5!(10-5)!} = \frac{10 \times 9 \times 8 \times 7 \times 6 \times \cancel{5 \times 4 \times 3 \times 2 \times 1}}{(5 \times 4 \times 3 \times 2 \times 1)\left(\cancel{5 \times 4 \times 3 \times 2 \times 1}\right)} = \frac{\cancel{10}^2 \times 9 \times \cancel{8}^2 \times 7 \times 6}{\cancel{5} \times \cancel{4} \times \cancel{3 \times 2} \times 1} = 252.$$

The number of combinations will always be smaller than the number of permutations. Use permutations when the arrangement or order is important and combinations when a group should only be counted once, no matter how it's arranged.

**CHECK POINT**

11. Find the number of combinations of 6 things taken 3 at a time.

12. Find the number of combinations of 7 things taken 4 at a time.

13. Find the number of combinations of 5 things taken 2 at a time.

14. How many different committees of 5 people can be chosen from a group of 12 people?

15. How many different ways can you order your ice cream sundae if you are going to choose 3 toppings from a list of 12 possibilities and the order in which you put toppings on does not matter?

# Relative Frequency

The frequency with which something happens is how often it happens, and that can be important information. It's probably good to know that I found money I had forgotten about in my coat pocket four times. But there's another piece of information that's missing. I found money four times, but was that four times last week? Four times last month? Four times last year? Four times in my life? The answer to that makes the "four times" information more useful.

The *relative frequency* of an event is a number between zero and one that compares the number of times something happens to the total number of things that happened. If I found money in my pocket four times last week, that's 4 days out of 7 days, or a relative frequency of $\frac{4}{7}$. If it was four times last year, the relative frequency is $\frac{4}{365}$.

> **DEFINITION**
>
> The **relative frequency** of an event is the ratio of the number of times it occurs to the total number of events observed.

Suppose you watched TV one evening and kept track of the commercials that aired in the few hours you watched. Here's your record.

## Commercials on Tuesday Night from 8 P.M. to 9 P.M.

| Type of Commercial | Number of Ads |
|---|---|
| Cars and trucks | 3 |
| Food and drink | 2 |
| Phones and tablets | 3 |
| Drugs and medicines | 3 |
| Security | 2 |
| Retail stores | 5 |
| Clothing and shoes | 1 |
| Other TV shows | 6 |

The chart tells you that more of the commercials you viewed were for other TV shows than anything else, but is 6 a lot? Six out of eight would be a lot, but you saw more than eight ads. Six out of a hundred would not be a lot, but you didn't sit through a hundred ads. How many did you watch? Add up the column. Then you can compare each of the counts to the total to get the relative frequency.

## Commercials on Tuesday Night from 8 P.M. to 9 P.M.

| Type of Commercial | Number of Ads | Relative Frequency |
|---|---|---|
| Cars and trucks | 3 | $\frac{3}{25} = 12\%$ |
| Food and drink | 2 | $\frac{2}{25} = 8\%$ |
| Phones and tablets | 3 | $\frac{3}{25} = 12\%$ |
| Drugs and medicines | 3 | $\frac{3}{25} = 12\%$ |
| Security | 2 | $\frac{2}{25} = 8\%$ |
| Retail stores | 5 | $\frac{5}{25} = 20\%$ |
| Clothing and shoes | 1 | $\frac{1}{25} = 4\%$ |
| Other TV shows | 6 | $\frac{6}{25} = 24\%$ |
| Total | 25 | |

Those relative frequencies are helpful in getting a sense of how the numbers relate to one another. You now know that almost one-fourth of the commercials you saw were for other TV shows, but also that one-fifth of them were for retail stores. You can also use the relative frequencies to make some predictions, or at least to talk about your expectations.

Suppose next Tuesday night you sit down to watch TV again, but you see 35 commercials instead of 25. You know that 20 percent of the ads you saw last time were for retail stores, so you can say that you'd expect about 7 of the 35 to be for stores, and 8 or 9 to be for other TV shows. You'd have a sense of what you're likely—and not likely—to see.

Relative frequency is a way to use your observations of events to get a sense of the *probability* of those events, or how likely they are to occur. If 12 percent of the commercials you saw on one evening were for phones and tablets, you would expect that on the next Tuesday night at the same hour, about 12 percent of the commercials will be for phones and tablets. You'd expect that the probability of seeing an ad for phones or tablets is about 12 percent. It's not guaranteed. The night you made your observations might have been an unusual night, or the night you're trying to predict might be special somehow. But you have a place to start talking about how likely something is.

**DEFINITION**

The **probability** of an event is the ratio of the number of ways the event can occur to the total number of events that can possibly occur.

# Theoretical Probability

Relative frequency is useful in estimating the probability of events, but you don't always have to do all that observing and recording. Sometimes you can make an estimate of the probability of some event based on things you already know (or can reasonably assume.) If you toss a coin, you know it has two sides—a head and a tail—and you assume that heads and tails are equally likely. As a result, you expect that the probability of heads is $\frac{1}{2}$ or 50 percent, and the probability of tails is also 50 percent. If you have a bag containing 10 marbles, 3 red, 4 blue, 2 green, and 1 yellow, and you are planning to draw out one marble without looking, you can reasonably expect that the probability that it will be red is $\frac{3}{10}$ or 30 percent, and the probability that it will be blue is $\frac{4}{10}$ or 40 percent. In the same way, the chance of drawing a green marble will be 20 percent, and the chance of a yellow will be 10 percent.

The probability of something happening is a number between zero and one that tells you how likely the event is. That number can be written as a fraction, a decimal, or a percent. If the probability is zero, the event is impossible. It can't happen. If the probability is one, it's absolutely certain to happen. It's a sure thing. If you draw one item out of that bag of ten marbles, the probability that it's a kitten is 0. The probability that it's a marble is 100 percent, or 1.

Most probabilities are somewhere in between zero and one, because most events are neither absolutely impossible nor absolutely certain. Probabilities are fractions, but you'll often hear them expressed as percentages (a 30 percent chance of rain) or as ratios (1 in 10 chance of such-and-such happening). The probability of an event is a fraction that compares the number of ways the event can happen to the total number of things that can happen.

$$P(E) = \frac{\text{\# of ways E can happen}}{\text{\# of things that can happen}}$$

If you take a well-shuffled deck of cards and pick one card at random, the probability of choosing an ace is $\frac{4}{52} = \frac{1}{13}$ because there are 4 aces out of 52 cards in the deck. The probability that the card will be a heart is $\frac{13}{52} = \frac{1}{4}$ because 13 of the 52 cards are hearts. The probability that you will choose the ace of hearts is $\frac{1}{52}$ because there is only one ace of hearts in the deck.

# Probability of Compound Events

Basic probability is pretty simple. The number of ways things can go right, over the number of ways things can happen. But life is complicated. Things don't always happen one at a time. In fact, they rarely happen one at a time. Usually there are at least two things to think about, but if you can handle them two at a time you can work your way through everything. It's all about whether you want one or the other, or you want both.

## Probabilities with "And"

Calculating the probability of a single card drawn at random from a standard deck being a 7 is fairly simple. There are four 7s out of 52 cards, so the probability is $\frac{4}{52}$ or $\frac{1}{13}$. Calculating the probability of a particular poker hand is much more complicated, because more cards are involved, and both the number of successes and the number of cards in the deck change as the deal goes on. That calculation begins, however, with finding the probability that you get two cards, two particular cards, one after the other. Let's say you want a heart and then a diamond.

That's drawing two cards, of course, not one. So how many ways are there to draw two cards from a deck of 52? That's the permutations of 52 things taken 2 at a time, or 52 × 51 = 2,652. And how many ways to get a heart and then a diamond? Go back to your basic counting principle. There are 13 ways to get a heart on the first draw, and there are 13 ways to get a diamond on the second draw. 13 × 13 = 169 ways to get a heart and then a diamond. The probability of a heart and then a diamond is 169 out of 2,652. Let's reduce that fraction.

$$\frac{169}{2,652} = \frac{\cancel{13} \times 13}{\underset{4}{\cancel{52}} \times 51} = \frac{13}{204}$$

Another way to think about this is that the probability you get a heart on the first draw is $\frac{13}{52}$, and the probability of a diamond on the second draw is $\frac{13}{51}$. (The denominator is only 51 because you've already taken one card out of the deck.) $\frac{\cancel{13}}{\underset{4}{\cancel{52}}} \times \frac{13}{51} = \frac{13}{204}$.

Any time you need to find the probability of this event *and* that one, you want to multiply the probability of the first event by the probability of the second event. Suppose you roll a die, record the number that comes up, then roll again and record the second result. What's the probability that both of them are even numbers? There are six possibilities for how the die can land, and three of them are even. The probability of two even numbers is $\frac{3}{6} \times \frac{3}{6} = \frac{9}{36} = \frac{1}{4}$.

Sometimes, when you look at the probability of two events occurring in sequence, the results of the first event have an effect on the probability of the second and sometimes it doesn't. When you roll the die twice, the first roll doesn't affect the second. They're what we call independent events. The result of the first roll has no effect on the probabilities for the second.

When you drew the two cards, on the second draw, there would only be 51 cards to choose from, and that changes the probability. Drawing two cards without replacement is an example of dependent events, because the result of the first draw changes the probability for the second draw. But if you choose one card at random from a deck, record what it is, then put it back in the deck and shuffle before you pull a second card, the probabilities for the second draw are the same as the first.

If two cards are drawn at random from a standard deck, with replacement—that is, the first card is drawn, recorded, and replaced before the second card is drawn—the probability of drawing two aces is $\frac{4}{52} \times \frac{4}{52} = \frac{1}{13} \times \frac{1}{13} = \frac{1}{169} \approx 0.6\%$. If two cards are drawn without replacement, however, the probability of drawing two aces is $\frac{4}{52} \times \frac{3}{51} = \frac{1}{13} \times \frac{1}{17} = \frac{1}{221} \approx 0.5\%$.

> **CHECK POINT**
>
> A card is drawn from a standard deck, recorded, and replaced in the deck. The deck is shuffled, and a second card is drawn.
>
> 16. Find the probability of drawing a heart and then a queen.
>
> 17. Find the probability of drawing a heart and then a heart.
>
> 18. Find the probability of drawing a black card and then a red card.
>
> Two cards are drawn at random from a standard deck, without replacing the first card.
>
> 19. Find the probability of drawing a king and a queen.
>
> 20. Find the probability of drawing two black cards.

## Probabilities with "Or"

When you look at the probability of event A and event B happening, you multiply the probability of A times the probability of B. Multiplying these two fractions, both less than one, gives a smaller number. A smaller probability makes sense because it's harder to have a success when success means having two different things happen than if success means having only one thing happen.

What if you want to find the probability of getting a five or a six when you roll a die? If you're willing to accept either number as a success, your chance of success increases—in this case, it doubles. The probability of rolling a five or a six is the probability of rolling a five plus the probability of rolling a six. $P(5 \text{ or } 6) = P(5) + P(6) = \frac{1}{6} + \frac{1}{6} = \frac{2}{6} = \frac{1}{3}$.

If you draw a card at random from a standard deck, what's the probability of drawing a queen or a 10? The probability of drawing a queen is $\frac{4}{52} = \frac{1}{13}$, and the probability of drawing a 10 is also $\frac{4}{52} = \frac{1}{13}$, so the probability of drawing a queen or a 10 is $\frac{4}{52} \times \frac{4}{52} = \frac{1}{13} \times \frac{1}{13} = \frac{2}{13}$.

Here's a question that sounds the same, but be very careful. Don't jump in without thinking it through. If you draw a card from a standard deck, what's the probability of drawing an ace or a black card? Start with the probability of drawing an ace. $P(ace) = \dfrac{4}{52} = \dfrac{1}{13}$. Then you want the probability of drawing a black card: $P(black) = \dfrac{26}{52} = \dfrac{1}{2}$. Before you add, think for a minute. There's a little problem, or more accurately, two little problems: the ace of spades and the ace of clubs. The probability of an ace was based on four aces, including the ace of spades and the ace of clubs. The probability of a black card was based on 26 black cards, including the ace of spades and the ace of clubs. The two black aces are being counted twice, so you'll have to make an adjustment for that.

$$P(\text{ace or black}) = P(\text{ace}) + P(\text{black}) - P(\text{black ace})$$
$$= \frac{4}{52} + \frac{26}{52} - \frac{2}{52}$$
$$= \frac{28}{52} = \frac{7}{13}$$

In general, the probability of A or B is the probability of A plus the probability of B minus the probability of A and B.

$$P(\text{A or B}) = P(\text{A}) + P(\text{B}) - P(\text{A and B})$$

You may wonder why this little adjustment wasn't necessary in the example about rolling the die. The answer is that when you roll a die, getting a five and getting a six are mutually exclusive outcomes. There is no way that you can get a five and a six on the same roll. When event A and event B are mutually exclusive, the probability of A and B is zero, so $P(\text{A or B}) = P(\text{A}) + P(\text{B}) - P(\text{A and B})$ just becomes $P(\text{A or B}) = P(\text{A}) + P(\text{B})$.

**CHECK POINT**

A bag contains 24 marbles. Six of the marbles in the bag are red, 4 are blue, 3 are white, 9 are green and 2 are yellow. One marble will be chosen at random.

21.  What is the probability that the marble chosen is red or blue?

22.  What is the probability that the chosen marble is yellow or blue?

23.  What is the probability that the marble is green or white?

24.  What is the probability that the chosen marble is yellow or red?

25.  What is the probability that the marble is red or orange?

## The Least You Need to Know

- To count quickly, take the number of options for each decision and multiply.
- The probability of an event is the number of ways it can happen over the total number of things that can happen.
- The probability of A and B is the probability of A times the probability of B.
- The probability of A or B is the probability of A plus the probability of B minus the probability of both.

# CHAPTER 19

# Graphs

Information is all around us. Whether in words or in numbers, it comes at us from the news, from the internet, from our studies, and from conversations. Having all the facts is a good thing, but only if you can make sense of all of it. One of the ways to help us make sense of all the information coming at us is to represent it in a well-designed chart or graph. Visual representations of data are often easier to understand than tables full of numbers.

In this chapter, you'll look at those charts and graphs from both sides. One of the most useful skills you can have is the ability to read a graph and pull the important information from it. Knowing what distinguishes a good graph from a not so good one is important, too. Sometimes the responsibility is on you to produce a graph, and whether you do that by hand or with the help of a computer, you need to know how to do it correctly. This chapter will look at several different types of graphs, how to read them, how to make them, and what kinds of information they best represent.

## In This Chapter

- Building bar graphs and histograms
- Constructing circle graphs
- Drawing line graphs

# Bar Graph

Bar graphs are just what they sound like: graphs that use bars to compare different quantities. The bars are usually vertical, but it is possible to make a horizontal bar graph, too. Each bar represents a different quantity, and the height (or horizontal length) of the bar corresponds to the size of the number.

## When to Use It

Bar graphs are usually used to compare the sizes of different categories of information. The labels on the horizontal axis are usually the names of the categories, and the heights of the bars show the sizes of each category. The bars may be in size order (in which case the graph is sometimes called a Pareto chart) or in some order that makes sense for the categories.

Bar graphs can also be used to show the change in some value over time, although a line graph is probably a better choice. In that case, the labels on the horizontal axis may be dates, months, years, or other time intervals.

## How to Draw It

If you wanted to show the relative sizes of the four classes in County High School, you might make a bar graph, with a bar to represent each class. If the freshman class has 583 students, the sophomore class 529, the junior class 492, and the senior class 451, you'd want to draw a horizontal axis with a space of equal size for each of the bars. Usually, you leave a little space between the bars. Then you'll need a vertical axis with a scale that lets you show the different class sizes. The largest class is 583, so your vertical axis needs to go at least that high, and it needs to be divided into small enough units that you can find 583, 529, 492 and 451. If you try to do a scale from 0 to 600 by tens, you may run into a problem.

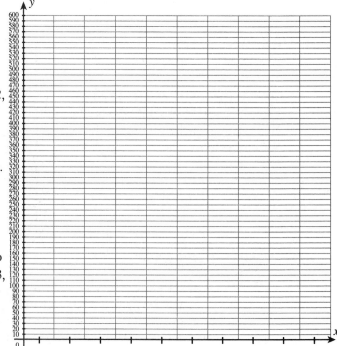

Either the vertical axis will become so crowded with numbers that you can't read them anymore, or the graph will become so big that it won't fit on the page. What can you do? One strategy is not to label every division. They're still there to help you to count the height of the bars, but the scale is easier to read.

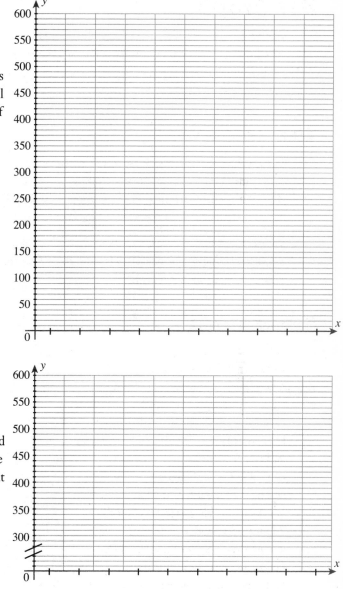

The other tactic you can use is what's called a broken scale. At the bottom of the vertical axis, you place a pair of slashes or a zigzag marker to show that you're skipping over some numbers, and then you begin the scale again after the break. This graph has a broken axis that picks up at 300.

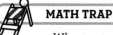 **MATH TRAP**

When a graph has a broken scale, the relative sizes of the bars can be distorted. If the scale jumps from zero to 500 in a short break, a bar that represents 700 could look almost twice the size of a bar that represents 600. Read carefully. Always check for broken scales and note how much of a jump they've taken.

So what does the bar graph for the four classes at County High School look like? You draw a bar for each class, making sure that the bars are all the same width, and bringing each up to a point on the scale that corresponds to the enrollment. You might want to label the bars with the actual values they represent, especially if the scale is hard to read.

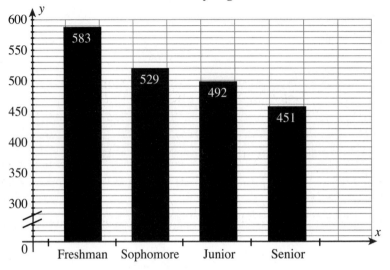

**Class Sizes at County High School**

## How to Read It

When you begin to read a bar graph, look first at the words around the graph. Is there a title to tell you what the graph is about? Are the scales labeled to tell you what they show and in what units they're measured? Then look at the actual bars. How high is each bar? You may need to estimate the quantities, if exact values are not given. Be sure to read scales carefully.

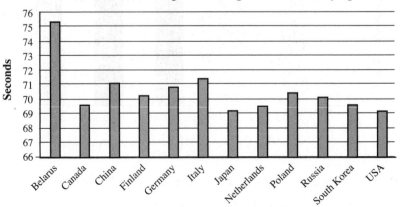

**Fastest Times for Speed Skating 2002 Winter Olympics**

This graph shows the best times in speed skating for athletes from different countries in the 2002 Winter Olympics. The horizontal axis shows the names of the countries. The vertical scale shows times in seconds. It's easy to see that the athletes from Belarus were significantly slower than those of the other countries represented. It's a little harder to find the fastest country, because several countries have times just over 69 seconds.

**CHECK POINT**

1. Draw a bar graph to represent the data below.

**Average Gasoline Usage in Hundreds of Gallons per Month for New England States**

| State | CT | ME | MA | NH | RI |
|-------|-----|-----|-----|-----|-----|
| Gasoline Usage | 4.3 | 5.4 | 4.4 | 5.4 | 4.1 |

2. Which state had the lowest average gasoline usage? Why might residents of that state use less gasoline?

The bar graph below summarizes the sales of various lunches offered in a school cafeteria.

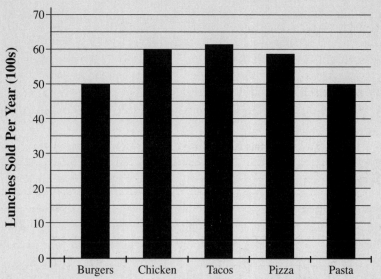

**Cafeteria Lunches**

3. Based upon this information, tacos outsell pasta by approximately how many lunches?

4. Approximately how many chicken lunches are sold per year?

5. Approximately how many burgers are sold per year?

# Histogram

There's another type of graph that, at first glance, may look a lot like a bar graph, but there are some key differences. The graph is called a histogram. It does use bars, but instead of representing different categories, this graph tells you how often a particular number or measurement showed up in the data. A histogram is an example of a frequency graph. It tells you how often, how frequently, a particular value occurs.

If you surveyed a group of people and asked how many pets they owned, you would likely get answers ranging from 0 to perhaps 5 or 6. It's also likely that there will be more than one person giving each of those answers. You could draw a histogram to show how many people gave each answer.

Notice that in the histogram the bars are touching, but in the bar graph there was space between them. That's because in a bar graph the categories on the horizontal axis are separate from one another. In the histogram, the horizontal axis has all the possible answers in numerical order, and every number on that axis is counted in one of the bars. Each bar picks up just where the last one left off, and no value on the horizontal axis is left out.

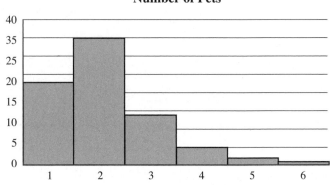

# Circle Graph

Circle graphs, sometimes called pie charts, are visual aids that show a circle broken into wedges, in much the way you would slice a pie or cake.

## When to Use It

Circle graphs are used to represent quantities as fractions of a whole. When you have several numbers that all combine to make up one thing, a circle graph may be a good way to show that. If you know that you have 8 seniors, 5 juniors, and 2 sophomores on your team, you might use a circle to represent the team, broken into three wedges. The size of each wedge corresponds to the fraction of the team that comes from each class.

**MATH TRAP**

Information given in percentages doesn't automatically go in a circle graph. A circle graph could show the percentages of the Town Council who come from each political party, but it can't show the percentage of Republicans in the Town Council compared to the percentage of Republicans in the state legislature. The percentages have to be parts of the same whole.

## How to Draw It

Drawing a circle graph seems easy. Make a circle, add a few lines to split it up, and attach a few labels. That will make a circle graph, but if you want a good circle graph, a really accurate circle graph, it takes a bit more work.

Suppose County High School made a list of how many of its students came from each of the towns in the county. Here's what they found out.

### Home Towns of County High Students

| Town | Number of Students |
|------|--------------------|
| Arlington | 274 |
| Buxton | 685 |
| Carleton | 137 |
| Dover | 411 |
| Eatontown | 548 |

To make an accurate circle graph, you first need to know what fraction of the high school enrollment comes from each town. To find that out, total up all the students, and divide the number from each town by the total of 2,055. Simplify the fractions.

### Home Towns of County High Students

| Town | Number of Students | Fraction of the School |
|------|--------------------|------------------------|
| Arlington | 274 | $\frac{274}{2055} = \frac{2}{15}$ |
| Buxton | 685 | $\frac{685}{2055} = \frac{1}{3}$ |
| Carleton | 137 | $\frac{137}{2055} = \frac{1}{15}$ |
| Dover | 411 | $\frac{411}{2055} = \frac{1}{5}$ |
| Eatontown | 548 | $\frac{548}{2055} = \frac{4}{15}$ |

Each town is going to be represented by a wedge of the circle, and the size of that wedge should be the same fraction of the circle. To make sure it's right, you need to find the number of degrees for each fraction of the circle and then get a protractor to measure the angles correctly. (Computer software can do a lot of the work for you, but it's good to know how to do it by hand.) Multiply the fraction of the school times 360 to find the number of degrees in each wedge, and then carefully divide the circle, using a protractor.

## Home Towns of County High Students

| Town | Number of Students | Fraction of the School | Number of degrees |
|------|-------------------|------------------------|-------------------|
| Arlington | 274 | $\frac{274}{2055} = \frac{2}{15}$ | $\frac{2}{15} \times 360° = 48°$ |
| Buxton | 685 | $\frac{685}{2055} = \frac{1}{3}$ | $\frac{1}{3} \times 360° = 120°$ |
| Carleton | 137 | $\frac{137}{2055} = \frac{1}{15}$ | $\frac{1}{15} \times 360° = 24°$ |
| Dover | 411 | $\frac{411}{2055} = \frac{1}{5}$ | $\frac{1}{5} \times 360° = 72°$ |
| Eatontown | 548 | $\frac{548}{2055} = \frac{4}{15}$ | $\frac{4}{15} \times 360° = 96°$ |

The final circle graph should look like this. Colors can help to emphasize the different wedges, but the drawing and the labels should be clear enough to make the information clear without colors.

## County High Enrollment by Town

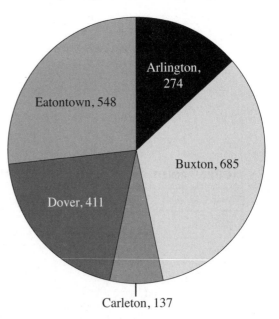

Arlington, 274

Eatontown, 548

Buxton, 685

Dover, 411

Carleton, 137

# How to Read It

When you begin to read a circle graph, the first thing you want to know is what the graph is talking about. Look for a title to tell you the general topic. This graph shows how the list of students who made the honor roll breaks down by class. Sometimes the graph will have a caption to explain more clearly what it's talking about. Next, look to the labels on the circle wedges. They should tell you what each of the wedges represents, and they may have values or percentages to tell you what part of the whole that wedge is. The graph below has four wedges, each labeled with the name of a class, and each has a percent that tells you what part of the honor roll comes from that class.

Sometimes your wedges are too small for labels or your labels are too long to fit neatly. Placing the label outside the circle near the wedge is the next best option, but if there are many small wedges, even that may be hard to do. In that case, the graph should have a legend. A legend is a small box near the graph that explains the code that you need to make sense of the graph. It should tell you what each color or pattern represents and probably include the percentages or values as well.

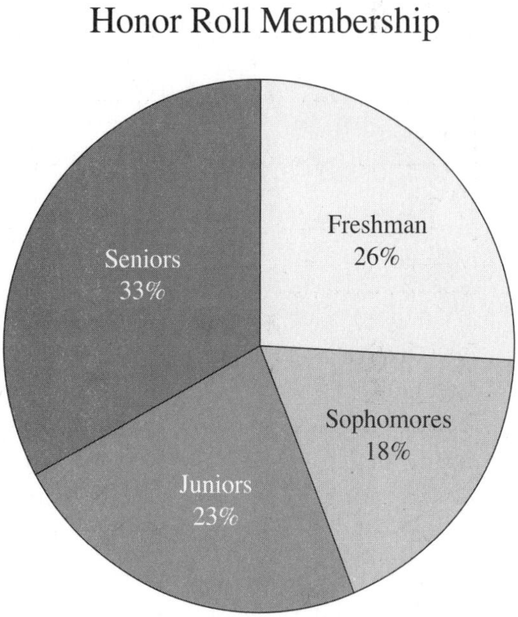

## Honor Roll Membership

- Seniors 33%
- Freshman 26%
- Sophomores 18%
- Juniors 23%

---

✏️ **CHECK POINT**

6. Draw a circle graph to represent the data below.

### Land Area of NYC by Borough

| Borough | Manhattan | The Bronx | Brooklyn | Queens | Staten Island |
|---------|-----------|-----------|----------|--------|---------------|
| Land area (square miles) | 23 | 42 | 71 | 109 | 58 |

7. The largest borough (by land area) is approximately how many times larger than the smallest?

This graph shows the enrollment in various arts electives last year.

8.  What percentage of the enrollment was in music courses?

9.  Which arts course had the largest enrollment?

10. Which two courses had the most similar enrollments?

11. If a total of 461 students signed up for arts electives, how many students took Art History?

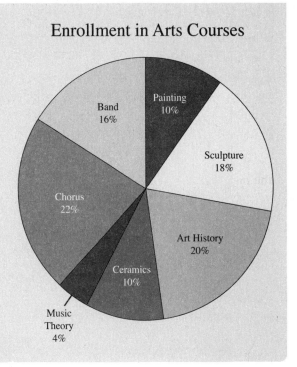

## Enrollment in Arts Courses

# Line Graph

A line graph is a graph that connects points with line segments. The lines help to clarify the pattern of change in the values the points represent.

## When to Use It

Line graphs are generally used to show the change in a quantity over time. A runner might record her times at each daily practice and create a line graph to help her see the pattern of improvement, or to see if there was a particular time when her times were slower or faster than usual.

## How to Draw It

To create a line graph, you need data that has been recorded. Each entry should include a date or time and a value. Let's suppose that runner recorded her times in seconds for the 200 meters every day in April. Here is her record for the first 10 days.

| Day  | 1     | 2     | 3     | 4     | 5     | 6     | 7     | 8     | 9     | 10    |
|------|-------|-------|-------|-------|-------|-------|-------|-------|-------|-------|
| Time | 22.71 | 22.68 | 22.50 | 22.43 | 22.37 | 22.19 | 22.04 | 21.97 | 21.91 | 21.89 |

The horizontal scale should show the day numbers, in this case up to 10, equally spaced along the axis. The vertical scale must go high enough to record all the times, so up to about 23 seconds. Then place a dot in line with each day's number at a height that corresponds to that day's time in seconds on the vertical axis. Finally, connect the dots, day 1 to day 2 to day 3 and so on, using straight line segments.

You may find that your first attempt is not satisfactory. If you labeled your vertical scale from 0 to 23 seconds, the graph may look very flat. The runner's times are only changing by a fraction of a second at a time.

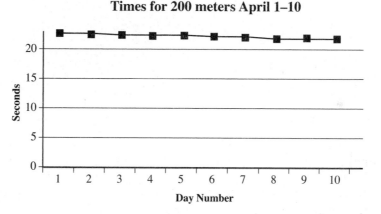

A broken scale is a good idea here. Here's what the same graph looks like if you use a broken scale, showing only 21.4 to 22.8 instead of 0 to 23. It's much easier to see the changes in the runner's times.

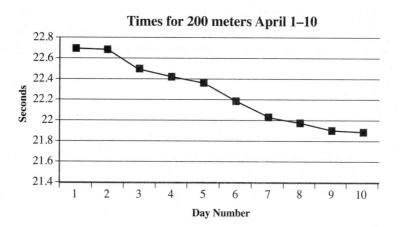

## How to Read It

Line graphs should have a title to tell you what information is being presented. Check that first, then examine the scales. The horizontal scale should show time, and it should be clear what the units of time are (years, months, days, etc.). This graph shows the average number of absences due to illness at County High in the months from January to June.

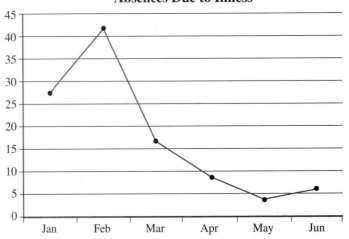

The range of numbers on the vertical scale and the units in which they are marked will depend on the information being presented, but take the time to read it and understand the units. This scale runs from 0 to 45, and the grid lines are every five units. Check for a broken scale and remember that a broken scale may distort your perception of size, just like in a bar graph, so examine the graph carefully. This scale is not broken, so there's no concern.

The line segments connecting the points will help you see overall patterns, but don't ignore the dots themselves. They are the actual observed and recorded data. The average number of absences due to illness at County High in May looks like 4. Connecting the points with lines is based on the assumption that the change from point to point is smooth and consistent, and that's not always true. Be very cautious about making any predictions or drawing any conclusions from what the line segments are doing. For example, the lines in this graph suggest that the sharpest decline in absentees took place from February to March, and that there were increases from January to February and from May to June. You wouldn't try to use the lines to say that there were 15 absentees on a certain date, however, because the line segments don't represent recorded data, only trends.

CHECK POINT

12. Draw a line graph to represent the data below.

### Esther's Spending on Restaurant Meals for the First 10 Days of August

| Date | 1-Aug | 2-Aug | 3-Aug | 4-Aug | 5-Aug | 6-Aug | 7-Aug | 8-Aug | 9-Aug | 10-Aug |
|------|-------|-------|-------|-------|-------|-------|-------|-------|-------|--------|
| Cost $ | 9.80 | 19.53 | 6.25 | 23.20 | 8.55 | 20.05 | 14.95 | 23.45 | 6.75 | 15.20 |

13. Is there any pattern to Esther's restaurant spending?

14. On which two days was Esther's spending most similar?

The graph below shows the sales of hotdogs at Jenny's Beach Bungalow over the course of the last year.

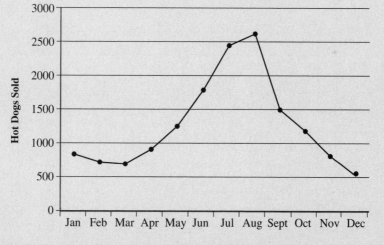

15. Over which period did sales have the greatest positive change?

16. Over which period did sales have the steepest drop?

17. Which autumn month had sales most similar to the number of hot dogs sold in February?

## The Least You Need to Know

- Bar graphs are used to compare the sizes of different quantities.
- A histogram shows how frequently a particular value occurs. Although they look similar, bar graphs and histograms are different types of graphs.
- Circle graphs show parts of a whole.
- A line graph tracks the change in one quantity over time.
- For any graph, always examine the information presented carefully. Pay attention to titles, units, and scale.

# CHAPTER 20

# Measures of Center and Spread

When researchers have a question, they perform lots of experiments or pose their question to many people and then they record all the results or responses. All that data can provide lots of information, but there may be so much data that it's difficult to make sense of it. It can turn into just a flood of numbers.

To make sense of information, you often need ways to summarize it and describe the patterns in it. Statistics are numbers that summarize collections of data or information. They help you to draw conclusions about the data.

In this chapter, you'll look at the statistics that give you the most important information. You'll find the measures of center, the numbers that tell you about the average or typical value. You'll also find the numbers that divide the data into groups, allowing you to compare different results. For the big picture, you'll talk about how spread out the data are, using a few simple numbers (and one not-so-simple one).

### In This Chapter

* Identifying mean, median, and mode
* Using percentiles and quartiles
* Measuring the spread of data

# The Centers

One of the ways you can represent a data set is by identifying a central tendency. A central tendency is the center value or a typical value of a particular data set. Identifying this value gives you a sense of the result you might get if you did the experiment again or asked the question again. Knowing this measure of center begins to give you a sense of the typical result, what you can expect. That expectation is approximate, of course. Every time you do the experiment or ask the question you will probably get a different value, but the more you do that, the better your sense of what to expect will become.

There are three different measures of center in common use. They're called the mean, the median, and the mode. The mean requires the most calculation, but unless you have an extremely large set of numbers, the calculation isn't difficult. The median only asks you to put numbers in order and the mode only requires you to count.

## Mean

The *mean* is the number most people think of when you say "average." It finds the center by balancing out the highs and lows of the numbers. The mean is found by adding all the data items and dividing by the number of items.

> **DEFINITION**
>
> The **mean** of a set of data is a measure of center found by adding all the data values and dividing by the number of values.

If you take three tests and earn grades of 84, 91, and 77, you find your average grade, or mean grade, by first adding up your three test scores. 84 + 91 + 77 = 252. Then you divide that total by 3 because it represents three tests. 252 ÷ 3 = 84. Your mean score is 84. The mean will not always be the same as one of the values, but it will be in the middle of the values. Your high score of 91 and your low score of 77 balance each other out, and the average ends up being the same as your middle score of 84.

If one or more of the values you're averaging are much higher or much lower than most of the group, the mean will be pulled toward that extreme value. Suppose you accidently copied that grade of 77 as just a 7. That would make that score significantly lower than your other grades. When you found the average, you'd get (84 + 91 + 7) ÷ 3 = 182 ÷ 3 = 60.67. That gigantic drop in your average is caused by that extremely low value.

Here's an example with a larger data set and larger numbers. Calculate carefully to find the mean amount of land in parks and wildlife areas in 2002 in the states shown below.

| Land in Rural Parks and Wildlife Areas 2002 (1,000 acres) | |
|---|---|
| Michigan | 1,436 |
| Wisconsin | 1,000 |
| Minnesota | 2,959 |
| Ohio | 372 |
| Indiana | 264 |
| Illinois | 432 |
| Iowa | 327 |
| Missouri | 649 |

First add the values given for each of the eight states. The total should be 7,439. Then divide by 8, because you added up the acreage for eight states. 7,439 ÷ 8 = 929.875. Keep in mind that the numbers you averaged are in thousands of acres, so the mean you calculated is not actually 929.875 acres, but 929.875 thousand acres or 929,875 acres.

**CHECK POINT**

Find the mean of each set of data

1. A = {2, 2, 2, 3, 3, 4, 4, 4, 4}

2. B = {34, 54, 78, 92, 101}

3. C = {3, 4, 5, 4, 7, 8, 9, 2, 10, 1}

4. D = {32, 34, 36, 38}

5. E = {2, 2, 3, 4, 5}

## Median

The *median* is the middle value when a set of data has been ordered from smallest to largest or largest to smallest. You've probably heard the word median used in other situations. In geometry, a median of a triangle connects a vertex to the midpoint of the opposite side. The strip of grass in the middle of a highway that divides the lanes moving in one direction from the lanes in the other direction is also called the median. Medians are always about the middle.

> **DEFINITION**
>
> The **median** of a set of data is the middle value when the data are ordered from smallest to largest. If the data set contains an even number of values, the median is the average of the two middle values.

To find a median, you just need to put the data in order and find the value that falls exactly in the middle of the list. If there is an even number of data points, and two numbers seem to be in the middle, the mean of those two numbers is the median. If the data set is very large, sorting and counting to find the middle can be tedious, and that's a good time to use a computer to help with the task. But for smaller data sets, finding the median is fairly simple.

Suppose a class of 10 students earned the grades below on an exam.

78, 59, 92, 82, 74, 97, 63, 75, 66, 88

To find the median grade, put the grades in order. Usually, people sort from low to high, but high to low will work, too.

59, 63, 66, 74, 75, 78, 82, 88, 92, 97

There are 10 grades, so the median will be the average of the fifth and sixth grades. Counting in from the low end, the fifth grade is 75 and the sixth is 78. The median is (75 + 78) ÷ 2 = 153 ÷ 2 = 76.5. The median grade is 76.5.

Earlier you found the mean number of acres of land in parks and wildlife areas for eight states in 2002. To find the median of the same data, first put the list in order by acreage.

| Land in Rural Parks and Wildlife Areas 2002 (1,000 acres) | |
|---|---|
| Minnesota | 2,959 |
| Michigan | 1,436 |
| Wisconsin | 1,000 |
| Missouri | 649 |
| Illinois | 432 |
| Ohio | 372 |
| Iowa | 327 |
| Indiana | 264 |

The median will be the average of the fourth and fifth values. (649 + 432) ÷ 2 = 1081 ÷ 2 = 540.5 thousand acres or 540,500. Remember that the mean was 929,875 acres. The mean is larger than the median because it's pulled toward the large acreage for Minnesota. The median doesn't get pulled in the same way, which is why statisticians say that the median is resistant.

**CHECK POINT**

Find the median of each set of data.

6. A = {2, 2, 2, 3, 3, 4, 4, 4, 4}

7. B = {34, 54, 78, 92, 101}

8. C = {3, 4, 5, 4, 7, 8, 9, 2, 10, 1}

9. D = {32, 34, 36, 38}

10. E = {2, 2, 3, 4, 5}

# Mode

Most people recognize the word *mode*, but not always from math. You've probably heard the phrase "à la mode" used in describing a dessert. It really doesn't mean "with ice cream." It actually means "according to the fashion." It was fashionable to add ice cream to a dessert, and doing so became quite common.

**DEFINITION**

The **mode** of a set of data is the value that occurs most frequently.

In statistics, the mode is the most fashionable value, the one that occurs most often. When you look at a set of data and see a value that is repeated frequently, that value may be the mode of the data set. If more than one value repeats, the one that repeats most often is the mode.

In the large data set below, you'll find several values that occur twice: 9.8, 10.8, 14.7, and 18.6. There is one value, 17.7, that occurs three times. The mode is the most common value, the one that occurs most frequently, so in this data set, the mode is 17.7.

| 6.9 | 8.7 | 9.7 | 10.7 | 11.5 | 13.3 | 14 | 14.7 | 17.3 | 18.6 |
|-----|-----|-----|------|------|------|----|------|------|------|
| 7.8 | 8.9 | 9.8 | 10.8 | 12.1 | 13.4 | 14.1 | 14.9 | 17.6 | 18.6 |
| 8.1 | 9 | 9.8 | 10.8 | 12.4 | 13.6 | 14.4 | 15.4 | 17.7 | 19 |
| 8.2 | 9.4 | 10.3 | 10.9 | 13 | 13.7 | 14.6 | 15.9 | 17.7 | 19.9 |
| 8.5 | 9.6 | 10.6 | 11.4 | 13.3 | 13.8 | 14.7 | 16.8 | 17.7 | 22.7 |

Not every data set will have a mode. Many are made up of values that are all different. And some will have more than one value that repeats the same number of times, so you can say they have more than one mode. Sometimes that's interesting information, but often it just means that there isn't a really common value.

The data set below shows the scores (out of 18) earned by 40 subjects in an experiment.

| Test Scores | | | | | | | |
|----|----|----|----|----|----|----|----|
| 14 | 14 | 8  | 9  | 13 | 11 | 13 | 9  |
| 10 | 9  | 11 | 15 | 13 | 12 | 13 | 9  |
| 16 | 12 | 15 | 14 | 8  | 12 | 17 | 15 |
| 15 | 16 | 11 | 16 | 14 | 16 | 13 | 12 |
| 17 | 7  | 13 | 8  | 16 | 14 | 14 | 10 |

To find the mode, it helps to sort the data, so that duplicate values are together and easier to count. Here's the same data sorted from low to high.

| Test Scores | | | | | | | |
|----|----|----|----|----|----|----|----|
| 7 | 9  | 11 | 12 | 13 | 14 | 15 | 16 |
| 8 | 9  | 11 | 12 | 13 | 14 | 15 | 16 |
| 8 | 9  | 11 | 13 | 13 | 14 | 15 | 16 |
| 8 | 10 | 12 | 13 | 14 | 14 | 16 | 17 |
| 9 | 10 | 12 | 13 | 14 | 15 | 16 | 17 |

With the data sorted, you can see that there are many repeated values, but you want to find the most common value. It turns out that this data set is bimodal. That means it has two modes: 13 and 14. Each of those scores occurs six times. The fact that they fall together is a strong indicator that these are typical values.

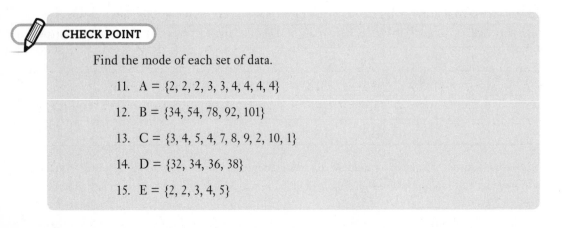

**CHECK POINT**

Find the mode of each set of data.

11. A = {2, 2, 2, 3, 3, 4, 4, 4, 4}

12. B = {34, 54, 78, 92, 101}

13. C = {3, 4, 5, 4, 7, 8, 9, 2, 10, 1}

14. D = {32, 34, 36, 38}

15. E = {2, 2, 3, 4, 5}

# The Separators

For small sets of data, you can take in the whole of the data set at once. The five test grades you earned over the course of a semester probably don't need to be broken up to help you understand them. But for larger sets, especially for very large collections of data, being able to divide the list up in organized ways can allow you to make comparisons that help you understand the important information.

You might want to compare numbers within the data set, or you might want to make comparisons between two related data sets. If a researcher was conducting a study on the cancer drugs, she might want to compare the test results for two subjects in the same drug trial, perhaps to see if gender or other factors changed the drug's effectiveness, or she might want to compare the results of the trials of two similar drugs to see which might be more effective.

## Quartiles and Percentiles

The most common ways of dividing up data sets are quartiles and percentiles. Quartiles divide the data into four equal parts, or quarters. Percentiles divide the data into 100 equal parts. Remember that percent means out of 100.

You could divide the data into any number of equal parts, and sometimes people do use other "tiles." Quintiles divide the data into five parts, for example. The quartiles and percentiles are the most common ones, however.

Percentiles are most useful for very large data sets. It would be impossible to divide a data set into 100 parts if it only contained 10 numbers. When there are hundreds or thousands of pieces of information, like the scores for the SAT test administered all across the country, percentiles can be a helpful way to compare values. If a score is at the 85th percentile, 85% of the scores are below it and 15% are above. If your score was right in the middle of all scores, so that you were at the median, you'd be at the 50th percentile. That means 50% of the scores are above you and 50%, or half, are below you.

For smaller sets of data, dividing into fewer parts makes more sense, and the most common is four parts, called quartiles. The first quartile is the value that has one-fourth of the data below it and three-fourths above. The second quartile has two-fourths, or half the data below and half above. The second quartile is never called the second quartile, however, because it's the median and always gets that name. The third quartile has 75 percent of the data below and 25 percent above. The first quartile, or Q1, the median, and the third quartile, Q3, divide the data set into four equal parts.

Suppose you had collected the data below and wanted to find the quartiles.

Number of books read last year: 16, 23, 13, 24, 25, 16, 17, 28, 19, 14, 12, 22, 13, 24, 15, 26, 27, 18, 29

Start by finding the median. Put the data in order, and find the middle value.

12, 13, 13, 14, 15, 16, 16, 17, 18, 19, 22, 23, 24, 24, 25, 26, 27, 28, 29

There are 19 numbers, so the tenth number is the median. That's 19. Then look at the numbers below the median, and find the middle value in that group. That value, 15, is the first quartile. Finally, look at the numbers above the median and find the middle value in that group. The number 25 is the third quartile.

## Boxplots

One of the ways in which quartiles are used is in the creation of a graph called a box and whisker plot, or a boxplot. It gets its name from its appearance. It's a box with a line poking out on each side, like a whisker. The whiskers reach to the lowest and the highest numbers in the data set, but the box shows the quartiles.

You have the data above on number of books read, and you know that Q1 = 15, the median is 19 and Q3 = 25. Notice that the smallest value is 12 and the largest is 29. Draw a scale and mark it to show at least 12 to 29. Draw a rectangle that reaches from 15 to 25. That's the box part of the box plot. From the lower end of the box, draw a line, a whisker, down to the minimum of 12. From the upper end, draw a whisker to the maximum of 29. Finally, add a divider to the rectangle at 19 to show where the median is.

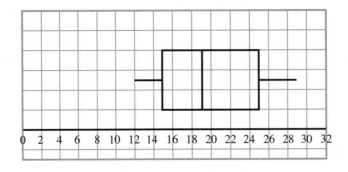

CHECK POINT

In questions 16–18, find the first quartile, median, and third quartile of each data set.

16.  A = {3, 4, 5, 4, 7, 8, 9, 2, 10, 17, 32, 34, 36, 38}

17.  B = {2, 2, 3, 3, 4, 4, 4, 34, 54, 78, 92, 101}

18.  C = {22, 83, 21, 49, 76, 64, 83, 29, 94, 19, 82, 28, 101}

19.  George and Harry take the same test. George's score places him at the $54^{th}$ percentile, and Harry's score is at the $43^{rd}$ percentile. Who did better?

20.  Draw a box plot for data set A in question 16.

# The Spread

It's good to know where the center of your data is. That tells you the average value. But only knowing the center of your data is like only knowing the center of a circle. You know where it is, but you don't really know what it looks like, because you don't know how big it is. You can't draw the circle until you know the center and the radius, and you don't have a good picture of your data until you know the center and the spread.

Measures of spread tell you whether all the numbers are clumped up close to the average or whether they're spread all over the place. If, over the course of the semester, you earned test scores of 69, 70, 73, 74, and 74, you'd have a mean score of 72 and a median score of 73. If you earned test scores of 43, 61, 73, 85 and 98, you'd also have a mean score of 72 and a median of 73, but the two sets of scores give very different pictures of how your semester went. Knowing how spread out the numbers are can also be important information.

## Range

The simplest measure of the spread is called the range. It's just the difference between the highest value and the lowest value. Those test scores of 69, 70, 73, 74, and 74 have a range of $74 - 69 = 5$, but the scores of 43, 61, 73, 85, and 98 have a range of $98 - 43 = 55$. The much larger range for the second set tells you that the numbers varied a great deal. The smaller range says that the numbers clumped up fairly close to the mean or median.

# Interquartile Range

The interquartile range, or IQR, is similar to the range, as you can tell from its name. The other part of the name, interquartile, means between the quartiles. To find the range, you subtract the minimum value from the maximum value. To find the interquartile range, you subtract the first quartile from the third quartile (Q3 – Q1).

Look back at the data about number of books read that you used to find quartiles.

Number of books read last year: 16, 23, 13, 24, 25, 16, 17, 28, 19, 14, 12, 22, 13, 24, 15, 26, 27, 18, 29

This data set has a minimum of 12 and a maximum of 29, for a range of 17. You found Q1 = 15, median = 19, and Q3 = 25. The interquartile range is Q3 – Q1 = 25 – 15 = 10.

The reason you sometimes want to use the interquartile range instead of the range is that the very high and very low values in your data often straggle far away from the other data. That exaggerates the range. The IQR cuts off those straggly parts but still gives you a sense of the spread.

# Standard Deviation

The third commonly used measure of spread is the one that's more complicated to find. It's called the standard deviation, and like the range and the IQR, the bigger it is, the more spread out the data are. The standard deviation tells you how much the other numbers in the data set vary from the average. For this reason, it is often paired up with the mean; for example, you might hear that a data set has a mean of 42 with a standard deviation of 3. A low standard deviation tells you that the numbers in the data set are close to the mean, while a high standard deviation indicates that the numbers in a data set are far from the mean.

Understanding the standard deviation is harder than understanding the range, but here's a way to think about what the mean and standard deviation tell you.

The mean tells you how to locate the center of the data set.

The IQR tells you where the middle 50 percent of the data is.

The range tells you where 100 percent of the data is.

The standard deviation breaks the range up into sections, letting you gauge how far from the mean another value is.

The standard deviation is like a ruler, measuring how from the center a value falls.

So what is the standard deviation? The deviation part refers to how far from the mean each number in the data set is. That's the simple piece. The standard part refers to the more complicated work that's done to avoid or eliminate things that could confuse the information.

Calculators and computers often do the work of finding a standard deviation, especially for large sets of numbers, but let's go through the steps once, just so you know what's going on. Let's use the test scores of 43, 61, 73, 85, and 98 from the earlier example about spread.

The first step in finding a standard deviation is to find the mean. You know from the earlier example that the mean of this data is 72. The next step is to subtract that mean from each number.

| Number | Number − Mean |
|--------|---------------|
| 43 | 43 − 72 = -29 |
| 61 | 61 − 72 = -11 |
| 73 | 73 − 72 = 1 |
| 85 | 85 − 72 = 13 |
| 98 | 98 − 72 = 26 |

The basic idea is to average these deviations, but if you add them up right now, the negative numbers and the positive numbers will cancel each other out. You could take the absolute value of each deviation, add the absolute values, and divide by the number of them you have. That would give you the *mean absolute deviation*. But the standard deviation calculation squares the deviations first. Like the absolute value, that makes everything positive.

| Number | Number − Mean | Squared |
|--------|---------------|---------|
| 43 | 43 − 72 = -29 | $(-29)^2 = 841$ |
| 61 | 61 − 72 = -11 | $(-11)^2 = 121$ |
| 73 | 73 − 72 = 1 | $1^2 = 1$ |
| 85 | 85 − 72 = 13 | $13^2 = 169$ |
| 98 | 98 − 72 = 26 | $26^2 = 676$ |

Now you add up all those squares and get 1,808. Next, divide. If your data set comes from the whole population of interest, divide by the number of values. If you're working with just a sample from a larger group, divide by one less than that. In this data set you have 5 values, so if this set is everyone's test, divide by 5. If it's only a sample of the test takers, divide by 4. $1,808 \div 5 = 361.6$ or $1,808 \div 4 = 452$. The last step is to undo the squaring by taking the square root of 361.6 or 452. $\sqrt{361.6} \approx 19.02$ or $\sqrt{452} \approx 21.26$. This set of test scores has a mean of 72 and a standard deviation of either 19.02 or 21.26. That's a pretty big standard deviation, either way, telling you those test scores are very spread out.

CHECK POINT

21. Find the range of {34, 54, 78, 92, 101}

22. Find the range of {3, 4, 5, 4, 7, 8, 9, 2, 10, 1}

23. Find the interquartile range of {3, 4, 5, 4, 7, 8, 9, 2, 10, 17, 32, 34, 36, 38}

24. Find the IQR of {2, 2, 2, 3, 3, 4, 4, 4, 4, 34, 54, 78, 92, 101}

25. Find the standard deviation of the test scores {69, 70, 73, 74, and 74}.

## The Least You Need to Know

- Measures of center and spread help you make sense of large amounts of data.
- The mean is the average of a set of numbers, the median is the middle number in a set of numbers, and the mode is most common number in a set of numbers.
- Numbers like quartiles and percentiles let you place one number within the group.
- Range, IQR, and standard deviation tell you how widely spread the information is.

# Extra Practice

Your journey through the world of numbers is coming to a close. I hope it was a good trip for you. Was it perfect? No? Well, I suppose it's rare that any trip is perfect. After any trip, no matter how wonderful, you tend to come home thinking of what you'll do next time.

Welcome to next time. If you want to make your math world tour better, you need to make your skills stronger. Did you have trouble with some of the topics? Practice. This part contains extra problems based on the material in each of the previous parts. You can use them all to keep your skills strong, or focus on the problems that correspond to topics you found difficult. Answers are provided in Appendix B, so you can check your work as you go along.

# CHAPTER 21

# Extra Practice

## Part I: Arithmetic

Use the number 8,472,019 to answer questions 1–5.

1. Name the digit in the hundreds place.

2. Name the digit in the hundred-thousands place.

3. Name the digit in the thousands place.

4. Name the digit in the tens place.

5. Name the digit in the millions place.

Write each number in scientific notation.

6. 7,300

7. 12,000

8. 903

9. 2,450,000

10. 691,000

Evaluate each expression.

11. $8 \times 3 + 5^2$

12. $27 - 8 \div 4 \times 10$

13. $3(7 - 4) - 4(7 - 3)$

14. $(56 - 8) \div 4 - 10$

15. $8 \times (3 + 5)^2$

16. $-19 + 39 \div -3$

17. $-5(-4 + -8) + 5(4 + 7)$

18. $(16 - 37) \div -7$

19. $20 \times -3 + (-7)^2$

20. $17 - 33 - 9(12 - 8)$

Find the following:

21. The prime factorization of 280.

22. The prime factorization of 5,005.

23. The GCF of 510 and 272.

24. The LCM of 91 and 119.

25. The GCF and the LCM of 42 and 90.

Perform each calculation.

26. $\dfrac{15}{28} \times 4\dfrac{2}{3}$

27. $7\dfrac{3}{5} \div 1\dfrac{9}{10}$

28. $\dfrac{3}{5} + \dfrac{4}{7}$

29. $4\dfrac{3}{35} - \dfrac{7}{15}$

30. $\dfrac{27}{50} - \dfrac{1}{2}$

31. $3.78 + 9.3$

32. $29.3 - 18.53$

33. $6.23 \times 2.04$

34. $4.374 \div 1.2$

35. $1.1 \div 0.88$

Use a proportion to solve each problem.

36. If 6 plants require 4.5 ounces of water, how much water will 27 of the same type of plant need?

37. The ratio of dark chocolate candies to milk chocolate candies in a particular mixture is 3 to 7. If you buy 5 pounds of the mixture, how many pounds of dark chocolate candies will it contain?

38. A certain rye bread recipe calls for 4 cups of wheat flour and 3 cups of rye flour. You have a 5-pound sack of rye flour, which measures out to 22.5 cups, and you'd like to use it all. How much wheat flour will you need?

39. Jeff is ordering food for a Super Bowl party. The caterer's brochure says that 6 gallons of chili will feed 40 people. If 100 people are expected at the party, how much chili should Jeff order?

40. Leslie designed a drapery panel that is made from 2 yards of print fabric and 5.5 yards of solid fabric. To cover all the windows in her living room, she will need 90 yards of fabric in total. How many yards of solid fabric should she order?

Solve.

41. What is 18% of 45?

42. Thirty-three is what percent of 110?

43. 38 is 16% of what number?

44. What is the percent decrease in the price of a rug that originally sold for $850 and is on sale for $629?

45. A mountain bike that originally sold for $600 increases in price to $675. What is the percent increase in the price?

# Part II: Algebra

Simplify each expression.

46. $3x - 7x + 12 - 2x + 5$

47. $4(5 - 3x) - 2(9x - 7)$

48. $\dfrac{35x - 23 + 15x - 12}{5}$

49. $\dfrac{2(5t - 1) + 8t + 2}{9t}$

50. $-7(5y - 2) + 8(6 - 5y)$

Solve each equation or inequality.

51. $7x - 12 = 5x + 4$

52. $3(2x - 9) = -3(x + 5)$

53. $11t - 49 = 2(7t - 5)$

54. $-4y - 6 > -11y + 1$

55. $11 - 3x \le 2x + 1$

Graph each equation.

56. $y = 6 - \dfrac{3}{5}x$

57. $x + y = 8$

58. $3x - y = 7$

59. $y = 5$

60. $x = -4$

# Part III: Geometry

Use the diagram to identify each of the following.

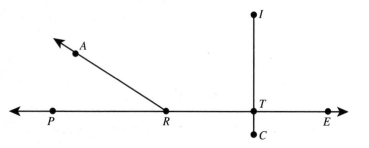

61. A ray

62. A line segment

63. An acute angle

64. An obtuse angle

65. A pair of vertical angles

Use the diagram to answer questions 66–70. $\overleftrightarrow{AI} \parallel \overleftrightarrow{BK}$ and $\overleftrightarrow{GL} \perp \overleftrightarrow{AI}$.

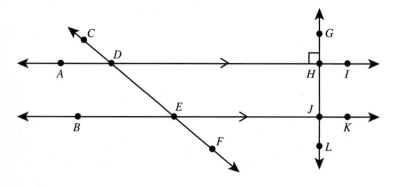

66. Name the angle that corresponds to $\angle HDE$.

67. $\angle CDH$ and $\angle BEF$ are a pair of _____ angles.

68. If m$\angle ADC = 20°$, find m$\angle DEJ$.

69. Find the measure of $\angle HJK$.

70. If m$\angle DEJ = 150°$, find the measure of $\angle HDE$.

Find the missing measurements.

71. In △*JAM*, *JA* = *AM*. If m∠*J* = 80°, find m∠*A*.

72. If *AR* = 7 cm, *RM* = 13 cm, and △*ARM* is isosceles, then *AM* = _____ cm or _____ cm.

73. △*LEG* is a scalene triangle. *LE* = 4 cm and *EG* = 17 cm. The length of $\overline{LG}$ must be at least _____ cm but not more than _____ cm.

74. In △*MAT*, *MA* = 4 inches, *AT* = 4 inches, and *MT* = 6 inches. Is △*MAT* a right triangle?

75. △*COW* is a right triangle with right angle ∠*O*. If *CW* = 65 cm, and *CO* = 25 cm, find the length of $\overline{OW}$.

76. If △*PAT* ≅ △*POT* and m∠*A* = 70°, then m∠*O* = _____.

77. △*DOG* ≅ △*CAT*, and *DO* = 28 in., *OG* = 34 in., and *DG* = 45 in. How long is $\overline{CT}$?

78. △*HOG* ~ △*PIG*, and ∠*OGH* and ∠*IGP* are vertical angles. If *HO* = 14 cm, *PI* = 21 cm, and *HG* = 18 cm, find *PG*.

79. △*RUG* ~ △*LAP*, and *RU* is 8 centimeters longer than *LA*. If *RG* = 14 cm and *LP* = 10 cm, find *RU*.

80. △*PLA* is a scalene triangle, and side $\overline{LA}$ is extended through *A* to point *Y* and beyond, to form exterior angle ∠*PAY*. If m∠*PAY* = 115° and m∠*L* = 75°, find m∠*P*.

Using the diagram, complete each sentence with *parallelogram, rectangle, rhombus,* or *square*.

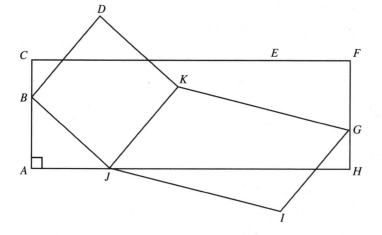

81. *ACFH* is a quadrilateral with $\overline{AC}$ ∥ $\overline{FH}$, $\overline{AH}$ ∥ $\overline{CF}$, and $\overline{CA}$ ⊥ $\overline{HA}$. *ACFH* is a _____.

82. *BJ* = *JK* = *KD* = *DB*. *BDKJ* is a _____.

83. $\overline{JK}$ ∥ $\overline{GI}$ and $\overline{JI}$ ∥ $\overline{KG}$. *JKGI* is a _____.

Find the measurement.

84. In rhombus *BDJK*, *BD* = 12 cm. In parallelogram *JKGI*, *GI* = _____ cm.

85. Find the area of a parallelogram with a base of 12 inches and a height of 5 inches.

86. *ABCD* is a trapezoid with $\overline{BC} \parallel \overline{AD}$. If m∠*A* = 58°, find m∠*B*.

87. In trapezoid *ABCD* with $\overline{BC} \parallel \overline{AD}$, *M* is the midpoint of $\overline{AB}$ and *N* is the midpoint of $\overline{CD}$. *BC* = 29 cm and *AD* = 41 cm. Find the length of $\overline{MN}$.

88. *PQRS* is a trapezoid with $\overline{QR} \parallel \overline{PS}$ and $\overline{PQ} \cong \overline{RS}$ If m∠*Q* = 119°, find m∠*S*.

89. Find the area of a trapezoid with height of 15 cm and bases 24 cm and 18 cm.

90. The area of a trapezoid is 375 cm², and the median is 15 cm long. Find the height.

91. Arc $\overset{\frown}{AB}$ measures 76°, and ∠*APB* has its vertex at point P on the circle. Find m∠*APB*.

92. Tangent $\overrightarrow{PQ}$ and tangent $\overrightarrow{PR}$ touch circle *O* at *Q* and *R* and intercept arcs of 59° and 301°. Find m∠*QPR*.

93. Chord $\overline{MN}$ intersects chord $\overline{ST}$ at point *X* inside circle *O*. Arc $\overset{\frown}{MS}$ measures 24° and arc $\overset{\frown}{NT}$ measures 52°. Find m∠*MXS*.

94. Find the circumference of a circle with radius 7 cm.

95. Find the area of a circle with diameter 24 cm.

Find the surface area of each of the figures described.

96. A square pyramid with side 8 cm and slant height 5 cm.

97. A triangular prism 11 cm high, with bases that are right triangles with legs 18 cm and 24 cm and hypotenuse 30 cm.

98. A cylinder with radius 15 inches and height 22 inches.

99. A cone with radius 5 inches, height 12 inches, and slant height 13 inches.

100. A sphere with a radius of 18 cm.

Find the volume of each of the figures described.

101. A rectangular prism with length 8 inches, width 5 inches, and height 14 inches.

102. A triangular pyramid 42 cm high with a base that is an equilateral triangle 26 cm on each edge, with an altitude of 22.5 cm.

103. A cylinder with a diameter of 20 cm and a height of 48 cm.

104. A cone with a radius of 50 inches and a height of 45 inches.

105. A sphere with a diameter of 60 cm.

Solve.

106. You stand 100 m from the base of a tall building and look up to the top of the building. You measure the angle between your line of sight and the horizontal as 72°. Use trigonometry to find the height of the building to the nearest foot.

107. The tip of the torch of the Statue of Liberty is 305.5 feet above the ground. If you stand on the edge of New York harbor and look to the tip of the torch and find that the angle is 3°, approximately how far from the Statue are you?

108. Jennifer is installing a bird feeder in her yard and wants to steady the pole it sits on by attaching wires to the pole and to stakes in the ground. She wants each wire to form the hypotenuse of a right triangle that is approximately isosceles. If each wire is 10 feet long, how far above the ground should she connect it to the feeder pole?

109. For safe use, a ladder should make an angle of about 75° with the ground. If you want to reach a point 30 feet above the ground and maintain the safe angle, how long should your ladder be?

110. What is the highest point on a wall that a 25-foot ladder can reach if it is placed at the safe 75° angle?

# Part IV: Probability and Statistics

For questions 111–115, assume that marbles are being drawn at random from a bag containing 5 red, 3 blue, and 2 yellow marbles.

111. If a single marble is drawn at random, what is the probability of drawing a blue marble?

112. If a single marble is drawn at random, what is the probability of drawing a red or a yellow marble?

113. A marble is drawn at random, recorded, and replaced in the bag, and then a second marble is drawn. What is the probability of drawing a yellow marble and then a blue marble?

114. A marble is drawn at random, recorded, and replaced in the bag, and then a second marble is drawn. What is the probability of drawing a blue marble and then a blue marble?

115. A marble is drawn, recorded, but not replaced in the bag. Another marble is drawn from the remaining marbles in the bag. What is the probability that both marbles are yellow?

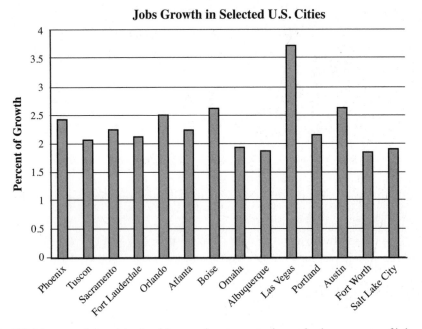

116. Which two of the cities in this sample appear to have the lowest rate of job growth?

117. How many of the cities in this sample appear to have job growth rates above 2.5%?

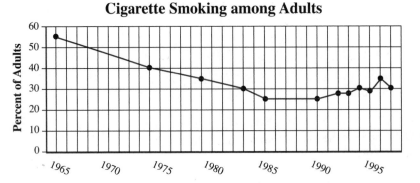

118. When did the percent of adults smoking cigarettes show an increase?

119. During which years did the percent of adults smoking cigarettes remain relatively constant?

Favorite Thanksgiving Food

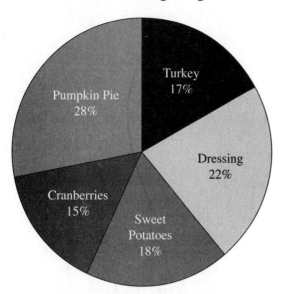

120.  What percent of people said their favorite food was a fruit or vegetable?

| 6 | 12 | 14 | 16 | 16 | 18 | 20 | 22 | 26 | 28 | 30 |
|---|---|---|---|---|---|---|---|---|---|---|

121.  Find the range.

122.  Find the median.

123.  Find the first and third quartiles.

124.  Find the mean.

125.  Find the standard deviation.

# Check Point Answers

## Chapter 1

1. In the number 3,492, the 9 is worth <u>9 tens</u>.

2. In the number 45,923,881, the 5 is worth <u>5 millions</u>.

3. In the number 842,691, the 6 is worth <u>6 hundreds</u>.

4. In the number 7,835,142, the 3 is worth <u>3 ten-thousands</u>.

5. In the number 7,835,142, the 7 is worth <u>7 millions</u>.

6. 79,038: seventy-nine thousand, thirty-eight

7. 84,153,402: eighty-four million, one hundred fifty-three thousand, four hundred two

8. "eight hundred thirty-two thousand, six hundred nine" = 832,609

9. "fourteen thousand, two hundred ninety-one" = 14,291

10. "twenty-nine million, five hundred three thousand, seven hundred eighty-two" = 29,503,782

11. $10,000 = 10^4$

12. $100,000,000,000 = 10^{11}$

13. $10^7 = 10,000,000$

14. $10^{12} = 1,000,000,000,000$

15. $10^5 = 100,000$

16. $59,400 = 5.94 \times 10^4$

17. $23,000,000 = 2.3 \times 10^7$

18. $5.8 \times 10^9 = 5,800,000,000$

19. $2.492 \times 10^{15} = 2,492,000,000,000,000$

20. $1.2 \times 10^{23} = 1.2 \times 10 \times 10^{22} = 12 \times 10^{22} > 9.8 \times 10^{22}$

21. $942 \approx 900$

22. $29,348 \approx 30,000$

23. $1,725,854 \approx 1,700,000$

24. $1,725,854 \approx 1,726,000$

25. $1,725,854 \approx 2,000,000$

# Chapter 2

1.  $48 + 86 = 134$

2.  $97 + 125 = 222$

3.  $638 + 842 = 1,480$

4.  $1,458 + 2,993 = 4,451$

5.  $12,477 + 8,394 = 29,871$

6.  $18 + 32 + 97 = 147$

7.  $91 + 74 + 139 = 304$

8.  $158 + 482 + 327 + 53 = 1,020$

9.  $71,864 + 34,745 + 9,326 = 115,935$

10. $9,865 + 7,671 + 8,328 + 1,245 + 3,439 = 30,548$

11. $596 - 312 = 284$

12. $874 - 598 = 276$

13. $1,058 - 897 = 161$

14. $5,403 - 3,781 = 1,622$

15. $14,672 - 5,839 = 8,833$

16. $100 - 62 = 38$

17. $250 - 183 = 67$

18. $500 - 29 = 471$

19. $400 - 285 = 115$

20. $850 - 319 = 531$

21. $462 \times 53 = 24,486$

22. $833 \times 172 = 143,276$

23. $1,005 \times 53 = 53,265$

24. $1,841 \times 947 = 1,743,427$

25. $2,864 \times 563 = 1,612,432$

26. $4,578 \div 42 = 109$

27. $3,496 \div 19 = 184$

28. $16,617 \div 29 = 573$

29. $681 \div 14 = 48$, remainder 9

30. $1,951 \div 35 = 55$, remainder 26

# Chapter 3

1.  $9 - 4 \times 2 = 9 - 8 = 1$

2.  $3^2 - 2 \times 4 + 1 = 9 - 8 + 1 = 2$

3.  $(2^3 - 5) \times 2 + 14 \div 7 - (5 + 1)$
    $= (8 - 5) \times 2 + 14 \div 7 - (5 + 1)$
    $= 3 \times 2 + 14 \div 7 - 6$
    $= 6 + 2 - 6$
    $= 2$

4.  $[(2^3 - 5) \times 2 + 14] \div 5 - 3 + 1$
    $= [(8 - 5) \times 2 + 14] \div 5 - 3 + 1$
    $= [3 \times 2 + 14] \div 5 - 3 + 1$
    $= [6 + 14] \div 5 - 3 + 1$
    $= 20 \div 5 - 3 + 1$
    $= 4 - 3 + 1$
    $= 1 + 1$
    $= 2$

5.  $[(32 - 2 \times 4) + 1]2 + [11 - 8 + 5(3 + 1)]$
    $= [(32 - 8) + 1]2 + [11 - 8 + 5(4)]$
    $= [(24) + 1]2 + [11 - 8 + 20]$
    $= [25]2 + [3 + 20]$
    $= 50 + 23 = 73$

6.  $2(35 + 14) = 70 + 28 = 98$

7.  $3(20 + 8) = 60 + 24 = 84$

8.  $7(100 - 2) = 700 - 14 = 686$

9.  $15(40 - 14) = 600 - 15 \times 14 = 600 - (150 + 60) = 600 - 210 = 390$

10. $250(1,000 - 400) = 250(600) = 150,000$

11. $|-19| = 19$

12. $|42| = 42$

13. $|0| = 0$

14. $|5 - 3| = |2| = 2$

15. $7 + |5 - 3| = 7 + |2| = 9$

16. $-15 + 25 = 10$

17. $19 + -12 = 7$

18. $-23 + 14 = -9$

19. $-58 + -22 = -80$

20. $147 + -200 = -53$

21. $-17 - 4 = -17 + (-4) = -21$

22. $39 - 24 = 39 + (-24) = 15$

23. $26 - -12 = 26 + 12 = 38$

24. $-83 - 37 = -83 + (-37) = -120$

25. $-48 - -32 = -48 + 32 = -16$

26. $-4 \times 30 = -120$

27. $8 \times -12 = -96$

28. $-7 \times 15 = -105$

29. $-11 \times -43 = 473$

30. $-250 \times 401 = -100,250$

31. $49 \div -7 = -7$

32. $-125 \div -15 = 8$, remainder 5

33. $-27 \div 9 = -3$

34. $120 \div -6 = -20$

35. $-981 \div -9 = 109$

# Chapter 4

1. 51 is composite. $51 = 17 \times 3$

2. 91 is composite. $91 = 13 \times 7$

3. 173 is prime.

4. 229 is prime.

5. 5,229 is composite. $5,229 = 3 \times 1,743 = 7 \times 747 = 9 \times 581 = 21 \times 249 = 63 \times 83$

6. $78 = 2 \times 3 \times 13$

7. $98 = 2 \times 7 \times 7$

8. $189 = 3 \times 3 \times 3 \times 7$

9. $255 = 3 \times 5 \times 17$

10. $512 = 2 \times 2 \times 2 \times 2 \times 2 \times 2 \times 2 \times 2 \times 2$

11. $200 = 2^3 \times 5^2$

12. $168 = 2^3 \times 3 \times 7$

13. $672 = 2^5 \times 3 \times 7$

14. $2,205 = 3^2 \times 5 \times 7^2$

15. $22,000 = 2^4 \times 5^3 \times 11$

16. GCF of 18 and 42 is 6.

17. GCF of 42 and 70 is 14.

18. GCF of 144 and 242 is 2.

19. GCF of 630 and 945 is 315.

20. GCF of 286 and 715 is 143.

21. LCM of 14 and 35 is 70.

22. LCM of 45 and 105 is 315.

23. LCM of 286 and 715 is 1,430.

24. LCM of 21 and 20 is 420.

25. LCM of 88 and 66 is 264.

# Chapter 5

1. $\dfrac{18}{5} = 3\dfrac{3}{5}$

2. $\dfrac{37}{4} = 9\dfrac{1}{4}$

3. $7\dfrac{1}{3} = \dfrac{22}{3}$

4. $12\dfrac{3}{4} = \dfrac{51}{4}$

5. $11\dfrac{7}{8} = \dfrac{95}{8}$

6. $\dfrac{1}{5} = \dfrac{8}{40}$

7. $\dfrac{3}{4} = \dfrac{21}{28}$

8. $\dfrac{5}{7} = \dfrac{105}{147}$

9. $\dfrac{35}{40} = \dfrac{7}{8}$

10. $\dfrac{63}{84} = \dfrac{3}{4}$

11. $\dfrac{2}{\cancel{3}} \times \dfrac{\overset{7}{\cancel{21}}}{\underset{25}{\cancel{50}}} = \dfrac{7}{25}$

12. $\dfrac{8}{15} \div \dfrac{2}{25} = \dfrac{\overset{4}{\cancel{8}}}{\underset{3}{\cancel{15}}} \times \dfrac{\overset{5}{\cancel{25}}}{\cancel{2}} = \dfrac{20}{3} = 6\dfrac{2}{3}$

13. $\dfrac{\overset{2}{\cancel{6}}}{\underset{7}{\cancel{49}}} \times \dfrac{\overset{2}{\cancel{14}}}{\underset{5}{\cancel{15}}} = \dfrac{4}{35}$

14. $4\dfrac{3}{8} \div 1\dfrac{7}{8} = \dfrac{35}{8} \div \dfrac{15}{8} = \dfrac{\overset{7}{\cancel{35}}}{\cancel{8}} \times \dfrac{\cancel{8}}{\underset{3}{\cancel{15}}} = \dfrac{7}{3} = 2\dfrac{1}{3}$

15. $5\dfrac{4}{7} \times 4\dfrac{2}{3} = \dfrac{\overset{13}{\cancel{39}}}{\cancel{7}} \times \dfrac{\overset{2}{\cancel{14}}}{\cancel{3}} = \dfrac{26}{1} = 26$

16. $\dfrac{4}{7} + \dfrac{3}{5} = \dfrac{20}{35} + \dfrac{21}{35} = \dfrac{41}{35} = 1\dfrac{6}{35}$

17. $\dfrac{8}{9} - \dfrac{1}{4} = \dfrac{32}{36} - \dfrac{9}{36} = \dfrac{23}{36}$

18. $\dfrac{2}{15} + 3\dfrac{1}{8} = \dfrac{16}{120} + 3\dfrac{15}{120} = 3\dfrac{31}{120}$

19. $7\dfrac{2}{3} + 9\dfrac{3}{4} = 7\dfrac{8}{12} + 9\dfrac{9}{12} = 16\dfrac{17}{12} = 17\dfrac{5}{12}$

20. $5\dfrac{1}{10} - 1\dfrac{1}{2} = 5\dfrac{1}{10} - 1\dfrac{5}{10} = 4\dfrac{11}{10} - 1\dfrac{5}{10} = 3\dfrac{6}{10} = 3\dfrac{3}{5}$

# Chapter 6

1. 9.003 is nine and three thousandths.

2. 82.4109 is eighty-two and four thousand one hundred nine ten-thousandths.

3. "forty-two hundredths" is 0.42.

4. "forty-two ten-thousandths" is 0.0042.

5. "three hundred twelve and nine hundred one thousandths" is 312.901.

6. $0.00001 = 10^{-5}$

7. $0.000000001 = 10^{-9}$

8. $0.000000000000001 = 10^{-15}$

9. $10^{-6} = 0.000001$

10. $10^{-10} = 0.0000000001$

11. $0.492 = 4.92 \times 10^{-1}$

12. $0.0000051 = 5.1 \times 10^{-6}$

13. $2.7 \times 10^{-5} = 0.000027$

14. $8.19 \times 10^{-7} = 0.000000819$

15. $5.302 \times 10^{-4} = 0.0005302$

16. $45.9 + 19.75 = 65.65$

17. $397.256 - 242.81 = 154.446$

18. $17{,}401.12 + 15{,}293.101 = 32{,}694.221$

19. $159.41006 - 143.0025 = 16.40756$

20. $1.00027 + 0.4587332 = 1.4590032$

21. $4.92 \times 1.5 = 7.380$ (or just 7.38)

22. $68.413 \times 0.15 = 10.26195$

23. $95.94 \div 7.8 = 12.3$

24. $461.44 \div 1.12 = 412$

25. $5{,}066.518 \div 8.6 = 589.13$

26. $\dfrac{4}{25} = \dfrac{16}{100} = 0.16$

27. $\dfrac{7}{9} = 0.777... = 0.\overline{7}$

28. $0.185 = \dfrac{185}{1{,}000} = \dfrac{37}{200}$

29. $0.\overline{123} = \dfrac{123}{999} = \dfrac{41}{333}$

30. $\dfrac{59}{8} = 7.375$

# Chapter 7

1. $5x + 3x = 32$
   $8x = 32$
   $x = 4$
   $5x = 20 \qquad 3x = 12$

   There are 12 boys in Math Club.

2. $7x + 1x = 40$
   $8x = 40$
   $x = 5$
   $7x = 35 \qquad 1x = 5$

   There were 5 hybrids sold.

3. $2x + 3x = 20$
   $5x = 20$
   $x = 4$
   $2x = 8 \qquad 3x = 12$

   The florist should use 12 white roses and 8 red roses.

4. $4x + 7x + 4x = 45$
   $15x = 45$
   $x = 3$
   $4x = 12 \qquad 7x = 21 \qquad 4x = 12$

   There are 21 tigers.

5. $21x + 20x + 9x = 900$
   $50x = 900$
   $x = 18$
   $21x = 378 \qquad 20x = 360 \qquad 9x = 162$

   There are 162 white balloons.

6. $\dfrac{2}{5} = \dfrac{x}{15}$
   $5x = 30$
   $x = 6$

7. $\dfrac{3}{7} = \dfrac{24}{x}$
   $3x = 168$
   $x = 56$

8. $\dfrac{x}{63} = \dfrac{15}{27}$
   $27x = 63 \times 15$
   $x = \dfrac{\overset{7}{\cancel{63}} \times \cancel{15}^{5}}{\cancel{27}_{\cancel{3}}} = 35$

9. $\dfrac{3}{x} = \dfrac{51}{68}$
   $51x = 3 \times 68$
   $x = \dfrac{\cancel{3} \times \cancel{68}^{4}}{\cancel{51}_{17}} = 4$

10. $\dfrac{3}{5} = \dfrac{x}{10}$

$5x = 30$

$x = 6$

11. 45 is 20% of 225.

$\dfrac{45}{x} = \dfrac{20}{100}$

$20x = 4500$

$x = 225$

12. 16 is 25% of 64.

$\dfrac{16}{64} = \dfrac{x}{100}$

$64x = 1600$

$x = 25$

13. 15% of 80 is 12.

$\dfrac{x}{80} = \dfrac{15}{100}$

$100x = 1200$

$x = 12$

14. 63 is 300% of 21.

$\dfrac{63}{21} = \dfrac{x}{100}$

$21x = 6300$

$x = 300$

15. 120% of 55 is 66.

$\dfrac{x}{55} = \dfrac{120}{100}$

$100x = 6600$

$x = 66$

16. $42\% = \dfrac{42}{100} = \dfrac{21}{50}$

17. $85.3\% = 0.853$

18. $\dfrac{5}{2} = 2.5 = 250\%$

19. $0.049 = 4.9\%$

20. $5.002 = 500.2\%$

21. $I = prt$

$I = 18000 \times 0.04 \times 5 = \$3,600$

The interest is \$3,600.

22. $I = prt$

$130 = 1000 \times r \times 2$

$r = \dfrac{130}{2000} = 6.5\%$

The interest rate is 6.5%.

23. $\text{tax} = 0.047 \times 175 = 8.225$

You will pay \$8.23 tax.

24. $\text{tip} = 0.20 \times \$35.84 = 7.168$

The tip should be approximately \$7.17.

25. A total of 8 people at \$22 per person is a bill of \$176. The tip will be $0.18 \times 176 = 31.68$ or \$31.68. Adding the tip to the bill brings the total to \$207.68. Dividing that total eight ways means that each person's share is \$25.96.

26. The increase is \$650 − \$500 = \$150.

$\dfrac{150}{500} = \dfrac{x}{100}$

$500x = 15000$

$x = 30$

The investment increased 30%.

27. The decrease is 7.5 − 6.75 = 0.75 minutes.

$\dfrac{0.75}{7.5} = \dfrac{x}{100}$

$7.5x = 75$

$x = 10$

Her time decreased 10%.

28. The change was an increase of $8.5 - 8 = 0.5$ pounds.

$$\frac{0.5}{8} = \frac{x}{100}$$

$$8x = 50$$

$$x = 6.25$$

The dog's weight increased 6.25%.

29. The decrease was $180 - 150 = 30$ pounds.

$$\frac{30}{180} = \frac{x}{100}$$

$$180x = 3000$$

$$x = 16\frac{2}{3}$$

His weight decreased $16\frac{2}{3}\%$.

30. The increase is $2 - 1.5 = 0.5$ quarts.

$$\frac{0.5}{1.5} = \frac{x}{100}$$

$$1.5x = 50$$

$$x = 33.\overline{3}$$

There was $33\frac{1}{3}\%$ more ice cream.

# Chapter 8

1. $\frac{x}{12} > 4$

2. $t - 19 = 22$

3. $n + (n + 3) = 154$

4. $y(y - 5) = 84$

5. $\frac{p}{p+1} = 2$

6. $5(-6a) = -30a$

7. $x(4x^2) = 4x^3$

8. $2y(-5y^3) = -10y^4$

9. $3t^2(-8t^5) = -24t^7$

10. $7z(3z + 5) = 21z^2 + 35z$

11. $\frac{-14t}{7} = -2t$

12. $\frac{11a^3}{a^2} = 11a,\ a \neq 0$

13. $\frac{18x^4}{6x^3} = 3x,\ x \neq 0$

14. $\frac{42y^7}{14y^4} = 3y^3,\ y \neq 0$

15. $\frac{12x^3 - 72x^2 + 32x}{4x} = \frac{12x^3}{4x} - \frac{72x^2}{4x} + \frac{32x}{4x} =$

$3x^2 - 18x + 8$

$x \neq 0$

# Chapter 9

1. $4x$ is a term.

2. $-12$ is a term.

3. $-2t^7$ is a term.

4. $\frac{6}{y}$ is not a term.

5. $\frac{a}{6}$ is a term.

6. $7y^2$ and $11y$ are unlike.

7. $3t^2$ and $5t^2$ are like.

8. $2x$ and $7x$ are like.

9. $-9a^2$ and $-15a^3$ are unlike.

10. $132x^3$ and $-83x^3$ are like.

11. $-4x + 9x = 5x$

12. $3a^2 - 2a^3$ is not possible.

13. $5xy + 6xy = 11xy$

14. $120xy^2 - 80xy^2 = 40xy^2$

15. $15z + 25x$ is not possible.

16. $6x(2x + 9) = 12x^2 + 54x$

17. $12 + 5(x + 1) = 12 + 5x + 5 = 5x + 17$

18. $6t^2(t - 3) - 2t^2 = 6t^3 - 18t^2 - 2t^2 = 6t^3 - 20t^2$

19. $5y(6y + 2) + 7y^2(4 - 12y) =$
    $30y^2 + 10y + 28y^2 - 84y^3 =$
    $-84y^3 + 58y^2 + 10y$

20. $8a(2b - 5) - 2b(a - 2) =$
    $16ab - 40a - 2ab + 4b =$
    $14ab - 40a + 4b$

21. $-3a^3 + 5a^2 - 3a + 12$ is degree 3.

22. $6 + 3b - 4b^2$ is degree 2.

23. $2t - 9 + 7t^2 + 4t^3$ is degree 3.

24. $11y - 7y^4 + 5y^2 - 3$ is degree 4.

25. $6 - 4x^2 + 3x$ is degree 2.

26. $a^3 + 10a^4 - 11a + 9 = 10a^4 + a^3 - 11a + 9$

27. $2b^3 - 9b + 12b^2 - 5 = 2b^3 + 12b^2 - 9b - 5$

28. $3k^4 + 8k^5 - 13k - 7 = 8k^5 + 3k^4 - 13k - 7$

29. $4 - 7m^2 + 14m^4 + 2m^3 =$
    $14m^4 + 2m^3 - 7m^2 + 4$

30. $5p - 3 + 6p^2 - 15p^3 = -15p^3 + 6p^2 + 5p - 3$

# Chapter 10

1. $x + 17 = 53$
   $x = 36$

2. $t - 11 = 46$
   $t = 57$

3. $-9a = 117$
   $a = -13$

4. $\dfrac{y}{6} = -14$
   $y = -84$

5. $x + 14 = -3$
   $x = -17$

6. $9x - 7 = -43$
   $9x = -36$
   $x = -4$

7. $\dfrac{y}{11} + 5 = 11$

   $\dfrac{y}{11} = 6$
   $y = 66$

8. $6 + 4x = 34$
   $4x = 28$
   $x = 7$

9. $\dfrac{t - 3}{5} = 12$
   $t - 3 = 60$
   $t = 63$

10. $7(x + 5) = 119$
    $7x + 35 = 119$
    $7x = 84$
    $x = 12$

11. $11x + 18 = 3x - 14$
    $8x + 18 = -4$
    $8x = -32$
    $x = -4$

12. $5(x+2)=40$

    $5x+10=40$

    $5x=30$

    $x=6$

13. $5(x-4)=7(x-6)$

    $5x-20=7x-42$

    $-2x-20=-42$

    $-2x=-22$

    $x=11$

14. $4(5x+3)+x=6(x+2)$

    $20x+12+x=6x+12$

    $21x+12=6x+12$

    $15x+12=12$

    $15x=0$

    $x=0$

15. $8(x-4)-16=10(x-7)$

    $8x-32-16=10x-70$

    $8x-48=10x-70$

    $-2x-48=-70$

    $-2x=-22$

    $x=11$

16. $2x-3=x-3$

    $x-3=-3$

    $x=0$

17. $6x-4=2(3x-2)$

    $6x-4=6x-4$

    Identity

18. $9x+11=3(3x+4)$

    $9x+11=9x+12$

    No solution

19. $5x+13=22-4x$

    $9x+13=22$

    $9x=9$

    $x=1$

20. $19-8x=4(5-2x)-1$

    $19-8x=20-8x-1$

    $19-8x=19-8x$

    Identity

21. $2x-5>13+4x$

    $-2x-5>13$

    $-2x>18$

    $x<-9$

22. $3x+2\leq 8x+22$

    $-5x+2\leq 22$

    $-5x\leq 20$

    $x\geq -4$

23. $12x+3<x+36$

    $11x+3<36$

    $11x<33$

    $x<3$

24. $2y-13\geq 4(2+y)$

    $2y-13\geq 8+4y$

    $-2y-13\geq 8$

    $-2y\geq 21$

    $y\leq -10.5$

25. $5x-10(x-1)>95$

    $5x-10x+10>95$

    $-5x+10>95$

    $-5x>85$

    $x<-17$

26.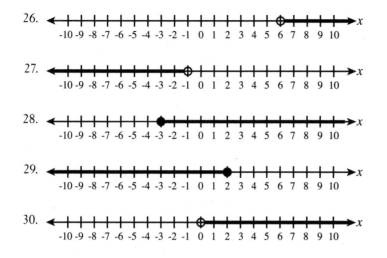

27.

28.

29.

30.

# Chapter 11

Answers for questions 1 through 5 are shown on one graph below.

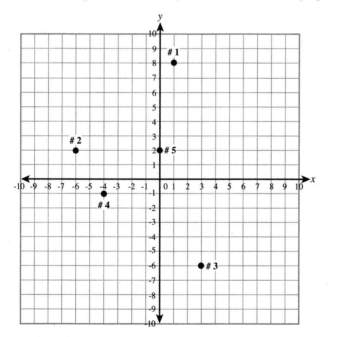

6. $x + y = 7$

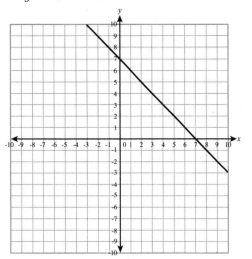

8. $y = 3x - 6$

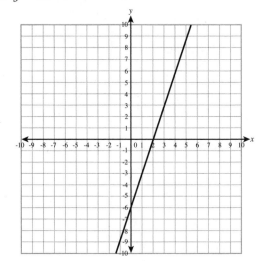

7. $2x - y = 3$

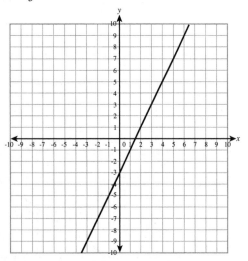

9. $y = 8 - 2x$

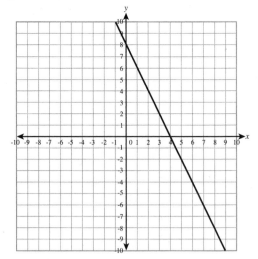

10. $y = \frac{2}{3}x - 1$

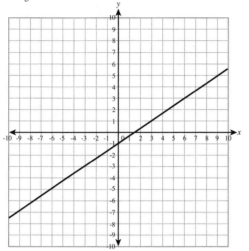

12. $6x + 2y = 12$ has intercepts $(0,6)$ and $(2,0)$.

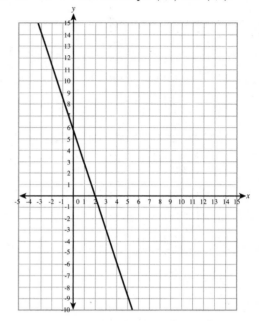

11. $x + y = 10$ has intercepts $(0,10)$ and $(10,0)$.

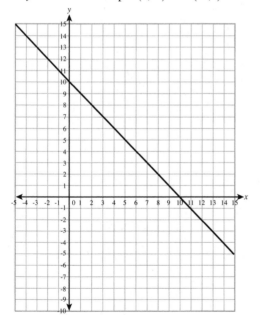

13. $2x - 3y = 9$ has intercepts $(0,-3)$ and $(4.5,0)$.

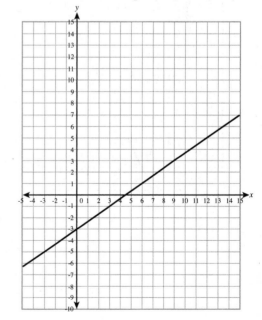

14. $x - 2y = 8$ has intercepts $(0,-4)$ and $(8,0)$.

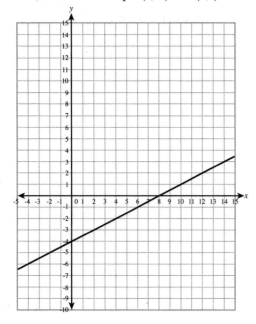

15. $6x + 2y = 18$ has intercepts $(0,9)$ and $(3,0)$.

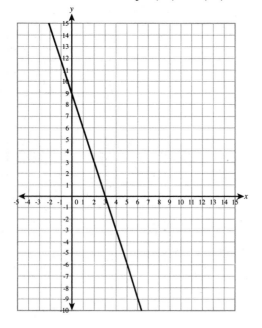

16. $m = \dfrac{y_2 - y_1}{x_2 - x_1} = \dfrac{5 - 2}{4 - 7} = \dfrac{3}{-3} = -1$

17. $m = \dfrac{y_2 - y_1}{x_2 - x_1} = \dfrac{-6 + 4}{9 - 6} = \dfrac{-2}{3}$

18. $m = \dfrac{y_2 - y_1}{x_2 - x_1} = \dfrac{7 - 6}{8 - 4} = \dfrac{1}{4}$

19. $m = \dfrac{y_2 - y_1}{x_2 - x_1} = \dfrac{-1 - 5}{5 + 5} = \dfrac{-6}{10} = -\dfrac{3}{5}$

20. $m = \dfrac{y_2 - y_1}{x_2 - x_1} = \dfrac{4 - 4}{8 - 3} = \dfrac{0}{5} = 0$

21. $y = -\dfrac{3}{4}x + 1$

$y$-intercept: $(0,1)$ slope: $-\dfrac{3}{4}$

22. $y = -4x + 6$

   $y$-intercept: (0,6) slope: -4

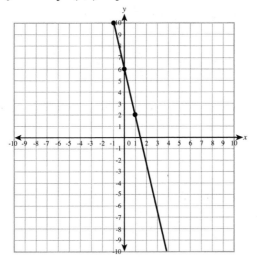

24. $2y = 5x - 6$

   $y$-intercept: (0,-3) slope: $\dfrac{5}{2}$

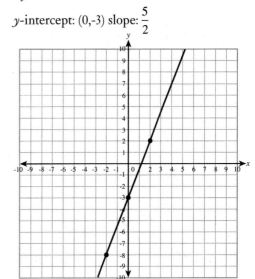

23. $y = -3x - 4$

   $y$-intercept: (0,-4) slope: -3

25. $y - 6 = 3x + 1$

   $y$-intercept: (0,7) slope: 3

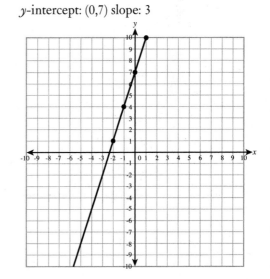

26. $y = -3$

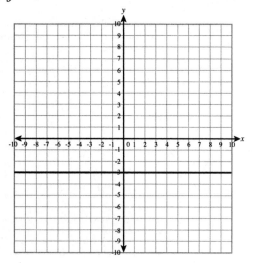

28. $y = 5$

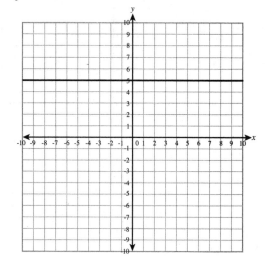

27. $x = 2$

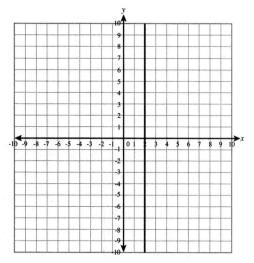

29. $x = -1$

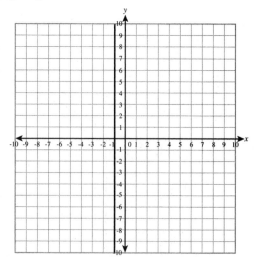

30. $y + 1 = 4$

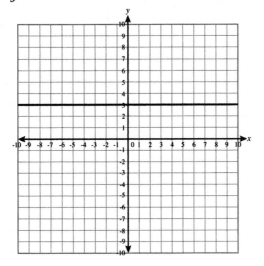

32. $y \leq 2x - 5$

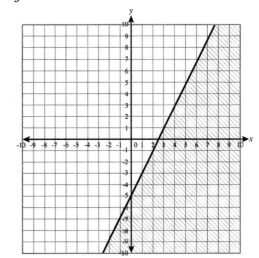

31. $y \geq \frac{1}{2}x + 1$

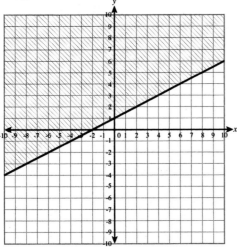

33. $y > 5x - 4$

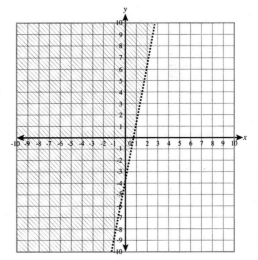

34. $y < x$

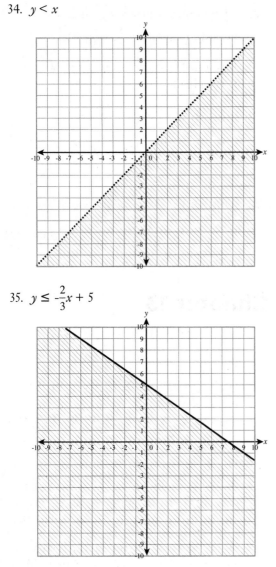

35. $y \leq -\frac{2}{3}x + 5$

# Chapter 12

Sample answers are shown for questions 1–5. Many answers are possible.

1. Line $PQ$

2. Ray $YZ$

3. Angle $\angle DEF$

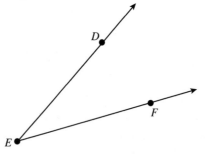

4. Rays $AB$ and $AC$

5. Angles $\angle PQR$ and $\angle RQT$

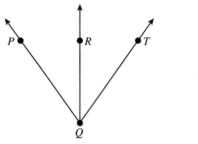

6. If $m\angle X = 174°$, then $\angle X$ is a(n) <u>obtuse</u> angle.

7. If $m\angle T = 38°$, then $\angle T$ is a(n) <u>acute</u> angle.

8. If $\angle X$ and $\angle Y$ are supplementary, and $m\angle X = 174°$, then $m\angle Y = \underline{6°}$.

9. If $\angle R$ and $\angle T$ are complementary, and $m\angle T = 38°$, then $m\angle R = \underline{52°}$.

10. Lines $\overrightarrow{PA}$ and $\overrightarrow{RT}$ intersect at point $Y$. If $m\angle PYR = 51°$, then $m\angle RYA = \underline{129°}$ and $m\angle TYA = \underline{51°}$.

11. $M$ is the midpoint of segment $\overline{PQ}$. If $PM = 3$ cm, $MQ = \underline{3}$ cm and $PQ = \underline{6}$ cm.

12. $H$ is the midpoint of $\overline{XY}$. If $XY = 28$ inches, then $XH = \underline{14}$ inches.

13. Ray $\overrightarrow{AH}$ bisects $\angle CAT$. If $m\angle CAT = 86°$, then $m\angle HAT = \underline{43°}$.

14. If $m\angle AXB = 27°$ and $m\angle BXC = 27°$, then $\overrightarrow{XB}$ bisects $\angle AXC$.

15. $m\angle PYQ = 13°$, $m\angle QYR = 12°$, $m\angle RYS = 5°$, and $m\angle SYT = 20°$. <u>True</u> or False: $\overrightarrow{YR}$ bisects $\angle PYT$. $m\angle PYR = 13° + 12° = 25°$. $m\angle RYT = 5° + 20° = 25°$.

16. $\angle PXY$ and $\angle XYT$ are a pair of <u>alternate interior</u> angles.

17. $\angle AXQ$ and $\angle XYT$ are a pair of <u>corresponding</u> angles.

18. If $m\angle XYT = 68°$, then $m\angle PXA = \underline{112°}$.

19. If $m\angle PXY = 107°$, then $m\angle RYB = \underline{107°}$.

20. If $\overrightarrow{AB} \perp \overrightarrow{PQ}$, then $m\angle XYR = \underline{90°}$.

21. Line $a$ is parallel to line $b$. Both have slopes of -3.

22. $\overrightarrow{RT}$ is perpendicular to $\overrightarrow{PQ}$. S negative reciprocals: $\dfrac{3}{5} \times -\dfrac{5}{3} =$

23. Line $p$ and line $q$ are neither parallel nor perpendicular.

24. $m_{\overline{XY}} = \dfrac{6+2}{7-4} = \dfrac{8}{3}$, $m_{\overline{WZ}} = \dfrac{4+8}{0-2} = {}^-6$. Line $\overleftrightarrow{XY}$ and line $\overleftrightarrow{WZ}$ are neither parallel nor perpendicular.

25. The line $3x - 2y = 12$ has slope $= \dfrac{3}{2}$ and the line $2x + 3y = 12$ has slope $= -\dfrac{2}{3}$. The lines are perpendicular.

# Chapter 13

1. $m\angle S = 180° - (48° + 102°) = 30°$

2. $m\angle TSQ = 48° + 102° = 150°$

3. $m\angle TSQ = 150°$, $m\angle TRP = 132°$, and $m\angle STN = 78°$
   The total of $m\angle TSQ + m\angle TRP + m\angle STN = 150° + 132° + 78° = 360°$.

4. Placidville is 43 miles from Aurora, and Aurora is 37 miles from Lake Grove. The distance from Placidville to Lake Grove is greater than <u>6</u> miles and less than <u>80</u> miles.

5. Gretchen lives 5 miles from the library and 2 miles from school. The distance from the library to school is between <u>3</u> miles and <u>7</u> miles.

6. $\triangle RST$ is isosceles with $RS = ST$. If $m\angle SRT = 39°$, then $m\angle STR = \underline{39°}$.

7. In right triangle $\triangle ABC$, $m\angle A = 90° - 19° = 71°$.

8. <u>False:</u> If $m\angle P = 17°$ and $m\angle Q = 25°$, $m\angle R = 180° - (17° + 25°) = 138°$. $\Delta PQR$ is an obtuse triangle, not an acute triangle.

9. If $m\angle P = 17°$ and $m\angle Q = 25°$, then the longest side of $\Delta PQR$ is side $\overline{PQ}$.

10. If the vertex angle of an isosceles triangle measures 94°, then the base angles measure <u>43° each</u>.

11. In $\Delta XYZ$, $\overline{XY} \perp \overline{YZ}$. If $XY = 15$ cm and $YZ = 20$ cm, <u>$XZ = 25$ cm</u>.

12. In $\Delta RST$, $\overline{ST} \perp \overline{RT}$. If $ST = 20$ inches and $RS = 52$ inches, <u>$RT = 48$ inches</u>.

13. In $\Delta PQR$, $\overline{PQ} \perp \overline{PR}$. If $PQ = PR = 3$ feet, <u>$QR = 3\sqrt{2} \approx 4.24$ feet</u>.

14. In $\Delta CAT$, $\overline{CA} \perp \overline{AT}$. If $CT = 8$ meters and $CA = 4$ meters, <u>$AT = 4\sqrt{3} \approx 6.93$ meters</u>.

15. In $\Delta DOG$, $\overline{DO} \perp \overline{OG}$. If $DO = 21$ cm and $DG = 35$ cm, <u>$OG = 28$ cm</u>.

16. $\Delta ABC$ is a 30°-60°-90° right triangle, with hypotenuse 8 cm long. <u>The length of the shorter leg is 4 cm</u>.

17. $\Delta RST$ is an isosceles right triangle with legs 5 inches long. <u>The length of the hypotenuse is $5\sqrt{2}$ inches</u>.

18. $\Delta ARM$ is a right triangle with $AR = 14$ meters, $RM = 28$ meters, and $AM = 14\sqrt{3}$ meters. <u>$m\angle M = 30°$</u>.

19. $\Delta LEG$ is a right triangle with $LE = EG$ and $LG = 7\sqrt{2}$ inches. <u>$m\angle G = 45°$</u>.

20. $\Delta OWL$ is an isosceles right triangle with $OW > OL$. <u>$\angle L$ is the right angle</u>.

21. $A = \dfrac{1}{2}bh = \dfrac{1}{2} \times 14 \times 7 = 49$ cm²

22. $A = \dfrac{1}{2}bh$

    $27 = \dfrac{1}{2}b \times 6$

    $27 = 3b$

    $b = 9$

    The base is 9 inches.

23. If the legs of the right triangle measure 20 cm and 48 cm, the hypotenuse is 52 cm.
    $P = 20 + 48 + 52 = 120$ cm.

24. In an equilateral triangle with a base $b$, the altitude will be $\dfrac{1}{2}b\sqrt{3}$, and the area is
    $A = \dfrac{1}{2}b\left(\dfrac{1}{2}b\sqrt{3}\right) = \dfrac{1}{4}b^2\sqrt{3}>$. If the area is $9\sqrt{3}$ square inches, $9\sqrt{3} = \dfrac{1}{4}b^2\sqrt{3}$, so $9 = \dfrac{1}{4}b^2$, $b^2 = 36$ and $b = 6$ inches. The perimeter is 18 inches.

25. The area of a right triangle with legs of 3 cm and 4 cm and hypotenuse of 5 cm is <u>6</u> square centimeters.

$$A = \frac{1}{2}bh$$
$$6 = \frac{1}{2} \times 5h$$
$$12 = 5h$$
$$h = 2.4$$

The altitude from the right angle to the hypotenuse is <u>2.4</u> centimeters long.

# Chapter 14

1. In quadrilateral $ABCD$, $\overline{AB} \parallel \overline{CD}$ and $\overline{BC} \parallel \overline{AD}$. $ABCD$ is a parallelogram because both pairs of opposite sides are parallel.

2. In quadrilateral $PQRS$, with diagonal $\overline{PR}$ $\angle QRP \cong \angle SPR$ and $\angle QPR \cong \angle SRP$. $PQRS$ is a parallelogram. The congruent alternate interior angles prove that both pairs of opposite sides are parallel.

3. In quadrilateral $FORK$, $\angle F \cong \angle K$ and $FO = RK$. There is sufficient information to say $FORK$ is a parallelogram.

4. In quadrilateral $LAMP$, with diagonals $\overline{LM}$ and $\overline{AP}$ intersecting at $S$, $\triangle ALS \cong \triangle PMS$ and $\triangle AMS \cong \triangle PLS$. The congruent triangles assure that both pairs of opposite sides are congruent, so $LAMP$ is a parallelogram.

5. In quadrilateral $ETRA$, with diagonals $\overline{ER}$ and $\overline{TA}$ intersecting at $X$, $TX = RX$ and $EX = AX$. There is not enough information to guarantee that $ETRA$ is a parallelogram.

6. In quadrilateral $FORT$, $\overline{FO} \perp \overline{OR}$, $\overline{OR} \perp \overline{RT}$ and $\overline{OR} \parallel \overline{FT}$. <u>$FORT$ is a rectangle</u>.

7. In quadrilateral $CAMP$, $CA = AM = MP = CP$ and $\overline{AM} \perp \overline{MP}$. <u>$CAMP$ is a square</u>.

8. In quadrilateral $VASE$, diagonals $\overline{VS}$ and $\overline{AE}$ are congruent, but sides $\overline{VA}$ and $\overline{AS}$ are not. $VASE$ is a rectangle.

9. In quadrilateral $SOAP$, $\overline{SO} \parallel \overline{AP}$ and $\overline{AO} \parallel \overline{SP}$. $SOAP$ is a parallelogram.

10. In quadrilateral $COLD$, diagonals $\overline{CL}$ and $\overline{OD}$ are perpendicular bisectors of one another, but they are not congruent. $COLD$ is a rhombus.

11. In trapezoid $ABCD$, $\overline{AC} \parallel \overline{BD}$ and $\overline{MN}$ is a median. $AC = 14$ cm and $BD = 30$ cm. <u>Median $\overline{MN}$ measures 22 cm</u>.

12. In trapezoid $FIVE$, $\overline{IV} \parallel \overline{FE}$ and $m\angle F = 59°$. $\underline{m\angle I = 121°}$

13. In trapezoid $TEAR$, $\overline{EA} \parallel \overline{TR}$ and $TE = AR$. If $\angle E = 107°$, $m\angle A = 107°$, and $m\angle R = 73°$.

14. In trapezoid $ZOID$, $\overline{ZD} \parallel \overline{OI}$, $m\angle Z = 83°$ and $m\angle I = 97°$. If $ZO = 4$ cm, $ID = 4$ cm, because $ZOID$ is an isosceles trapezoid.

15. In trapezoid $PQRT$, $\overline{PT} \parallel \overline{QR}$ and $\overline{MN}$ is a median. $\dfrac{PT + QR}{2} = MN$.
    If $MN = 17$ inches and $PT = 21$ inches,

$$\frac{21 + QR}{2} = 17$$
$$21 + QR = 34$$
$$QR = 13 \text{ inches.}$$

16. For a square with a side of 17 cm, perimeter is 68 cm and area is 289 cm².

17. For a rectangle 18 inches long and 9 inches wide, perimeter is 54 inches and area is 162 square inches.

18. For parallelogram $ABCD$, $AB = CD = 7$ inches, $AC = BD = 21$ inches, and the height from $B$ perpendicular to $\overline{AC}$ and $\overline{AD}$ 3 inches, perimeter is 56 inches, and area is 63 square inches.

19. For a rhombus with sides 5 inches long and diagonals that measure 6 inches and 8 inches, perimeter is 20 inches and area is 5 square inches.

$$A = \frac{1}{2}d_1d_2 = \frac{1}{2}\ 6 \cdot 8 = 24$$

20. If the area of a parallelogram with a height of 48 cm is 3,600 square centimeters, the base to which that altitude is drawn (and the opposite side) must measure 75 cm. If the perimeter is 250 cm, the other two sides each measure 50 cm.

21. The number of diagonals in an octagon is $\dfrac{8 \cdot 5}{2} = 20$.

22. The total of the measures of all the interior angles in a nonagon is $180° (9 - 2) = 1260°$.

23. If a hexagon is regular, the total of the measures of all the interior angles is $180° (6 - 2) = 720°$, and the measure of any one of its interior angles is $\dfrac{720°}{6} = 120°$.

24. If the interior angles of a polygon add up to 900°, then $900° = 180°(n - 2)$, and $n - 2 = 5$. The polygon has 7 sides.

25. If a polygon has a total of 119 possible diagonals, $\dfrac{n(n-3)}{2} = 119$ so $n(n-3) = 238$. The factors of 238 are 2 × 119, 7 × 34, and 14 × 17. The last pair differ by 3, so $n = 17$.

26. The area of a regular pentagon with sides 8 cm long and an apothem 5 cm long is
$$A = \frac{1}{2}ap = \frac{1}{2} \cdot 5 \cdot 40 = 100 \text{ cm}^2.$$

27. The area of an octagon with a perimeter of 40 inches and an apothem of 5 inches is
    $A = \frac{1}{2}ap = \frac{1}{2} \cdot 5 \cdot 40 = 100$ square inches.

28. The area of a regular decagon in which each of the 10 sides measure 2 meters and the apothem is
    1.5 meters is $A = \frac{1}{2}ap = \frac{1}{2} \cdot 1.5 \cdot 20 = 15$ square meters.

29. $A = \frac{1}{2}ap$

    $84 = \frac{1}{2}a \cdot 42$

    $84 = 21a$

    $a = 4$

    If the perimeter of a regular hexagon is 42 inches and its area is 84 square inches, its apothem is
    4 inches.

30. $A = \frac{1}{2}ap$

    $1080 = \frac{1}{2} \cdot 18 \cdot p$

    $1080 = 9p$

    $p = 120$

    A regular pentagon with an area of 1,080 square inches and an apothem of 18 inches has a
    perimeter of 120 inches. Each side is 24 inches.

# Chapter 15

1. An arc less than half a circle is a <u>minor</u> arc.

2. The distance from the center point to any point on the circle is called the <u>radius.</u>

3. Two circles with the same center are <u>concentric</u> circles.

4. If two circles touch each other at just one point, the circles are <u>tangent</u>.

5. An arc that is exactly half the circle is called a <u>semicircle</u>.

6. $m\angle TOP = 48°$

7. $m\angle MAN = \frac{1}{2}(78°) = 39°$

8. $m\angle SIT = \frac{1}{2}(36°) = 18°$

9. $\angle ITS$ is inscribed in a semicircle. $m\angle ITS = 90°$.

10. $m\angle GAL = \dfrac{56 + 82}{2} = 69°$ and

$m\angle PAL = 180 - 69 = 111.$

11. $m\angle APE = \dfrac{1}{2}(220 - 140) = \dfrac{1}{2}(80) = 40°$

12. $m\angle FAT = \dfrac{1}{2}(52 - 12) = \dfrac{1}{2}(40) = 20°$

13. $\angle RAC$ intercepts an arc of $360 - 72 = 288°$.
    $m\angle RAC = \dfrac{1}{2}(288) = 144°$.

14. Let $x$ = the measure of arc $\overparen{QR}$.

    $m\angle P = \dfrac{1}{2}\left(\overparen{QS} - \overparen{QR}\right)$

    $15° = \dfrac{1}{2}(160 - x)$

    $30 = 160 - x$

    $x = 130°$

15. $m\angle P = \dfrac{1}{2}(2x - x) = \dfrac{x}{2}$, but $2x + x = 360$,

    so $x = 120$ and $m\angle P = 60°$.

16. The area of a circle with a radius of 9 cm is $81\pi$ cm².

17. The circumference of a circle with a diameter of 12 inches is $12\pi$ inches.

18. The area of a circle with a diameter of 32 cm is $16^2\pi = 256\pi$ cm².

19. The radius of a circle with an area of $36\pi$ square meters is 6 meters, its diameter is 12 meters, and its circumference is $12\pi$ meters.

20. The diameter of a circle with a circumference of $24\pi$ feet is 24 feet, its radius is 12 feet, and its area is $144\pi$ square feet.

21. The equation of a circle with its center at the origin and a radius of 3 units is $x^2 + y^2 = 9$.

22.

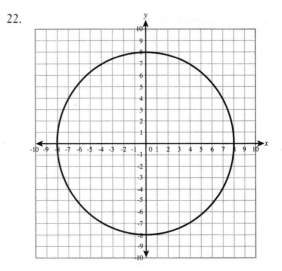

23. The center of the circle $(x - 8)^2 + (x - 3)^2 = 49$ is the point $(8,3)$ and radius is 7.

24. The equation of a circle with center $(4,9)$ and radius of 2 units is $(x - 4)^2 + (y - 9)^2 = 4$

25.

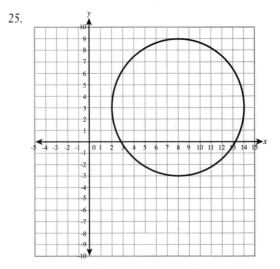

Circle of radius 6 centered at $(8,3)$

# Chapter 16

1.  SA = 2(15 × 24) + 2(15 × 10) + 2(24 × 10) = 1,500 cm²

2.  SA = $2(\frac{1}{2} \times 5 \times 12) + 5(8 + 12 + 13)$ = 225 square inches

3.  SA = 2 × 65 + 42 × 30 = 1,390 cm²

4.  SA = 2 × 387 + 5 × 15 × 4 = 1,074 square inches

5.  SA = 6 × 17² = 1,734

6.  $V = 7^3 = 343$ cubic inches

7.  $V = 12 \times 21 \times 15 = 3,780$ cm³

8.  $V = \frac{1}{2} \times 3 \times 4 \times 6 = 36$ cubic inches

9.  $V = 387 \times 8 = 3,096$ cubic inches

10. $V = 65 \times 50 = 3,250$ cm³

11. SA = $4^2 + \frac{1}{2}(16 \times 5)$ = 16 + 40 = 56 square inches

12. SA = $62.4 + \frac{1}{2}(36 \times 10)$ = 62.4 + 180 = 242.4 cm²

13. SA = $172 + \frac{1}{2}(50 \times 18)$ = 172 + 450 = 622 cm²

14. SA = $260 + \frac{1}{2}(60 \times 10)$ = 260 + 300 = 560 square inches

15. SA = $10^2 + \frac{1}{2}(40 \times 13)$ = 100 + 260 = 360 square inches

16. If the slant height is 13 inches and half the side is 5 inches, the height is 12 inches. $V = \frac{1}{3} \ 10^2 \cdot 12 = 400$ cubic inches.

17. If the slant height is 5 inches and half the side is 2 inches, the height is $\sqrt{21} \approx 4.58$ inches. $V = \frac{1}{3} \ 4^2 \cdot 4.58$ cubic inches.

18. The base of the pyramid is an equilateral triangle with a side of 12 cm and an area of 62.4 square centimeters. The area is half the apothem times the perimeter, so $62.4 = \frac{1}{2}a(3 \times 12)$ and the apothem is $a \approx 3.47$. Use the Pythagorean Theorem with the apothem and slant height to find the height. $a^2 + h^2 = l^2$ becomes $(3.47)^2 + h^2 = 10^2$ and $h \approx 9.38$. The height is approximately 9.38, and $V = \frac{1}{3} \times (62.4) \times (9.38) \approx 195.104$ cubic centimeters.

19. Use the area of the pentagon and its perimeter to find the apothem. $172 = \frac{1}{2}a(50)$ so $a \approx 6.88$. Use the Pythagorean Theorem to find the height. $a^2 + h^2 = l^2$ so $(6.88)^2 + h^2 = (18)^2$ and $h \approx 16.63$. $V = \frac{1}{3} \times (172) \times (16.63) \approx 953.45$ cubic centimeters.

20. The regular hexagon that forms the base has a perimeter of 60 inches and an area of 260 square inches, so use the formula $A = \frac{1}{2}aP$ to find the apothem. $260 = \frac{1}{2}a(60)$ means that the apothem is $8\frac{2}{3}$ inches long. Use the Pythagorean Theorem with the apothem and the slant height to find the height. $a^2 + h^2 = l^2$ so $\left(8\frac{2}{3}\right)^2 + h^2 = (10)^2$ and $h \approx 4.99$ inches. $V = \frac{1}{3} \times (260) \times (4.99) \approx 432.37$ cubic inches.

21. $h = 14$ cm, $r = 5$ cm, SA = $2 \cdot 5^2 \cdot \pi + 2 \cdot \pi \cdot 5 \cdot 14$ = $190\pi$ cm², $V = \pi \cdot 5^2 \cdot 14 = 350\pi$ cm³.

22. $b = 8$ inches, $d = 6$ inches, $r = 3$ inches, SA $= 2 \cdot 3^2 \cdot \pi + 2 \cdot \pi \cdot 3 \cdot 8 = 66\pi$ square inches, $V = \pi \cdot 3^2 \cdot 8 = 72\pi$ cubic inches.

23. $b = 2$ m, $C = 2\pi$ m, $d = 2$ m, $r = 1$ m, SA $= 2 \cdot 1^2 \cdot \pi + 2 \cdot \pi \cdot 1 \cdot 2 = 6\pi$ m², $V = \pi \cdot 1^2 \cdot 2 = 2\pi$ m³.

24. $b = 82$ cm, $d = 90$ cm, $r = 45$ cm, SA $= 2 \cdot 45^2 \cdot \pi + 2 \cdot \pi \cdot 45 \cdot 82 = 11,430\pi$ cm², $V = \pi \cdot 45^2 \cdot 82 = 166,050\pi$ cm³.

25. $b = 20$ inches, $C = 20\pi$ inches, $d = 20$ inches, $r = 10$ inches, SA $= 2 \cdot 10^2 \cdot \pi + 2 \cdot \pi \cdot 10 \cdot 20 = 600\pi$ square inches, $V = \pi \cdot 10^2 \cdot 20 = 2,000\pi$ cubic inches.

26. $r = 10$ cm, $b = 24$ cm, $l = 26$ cm, SA $= \pi \cdot 10^2 + \pi \cdot 10 \cdot 26 = 360\pi$ cm², $V = \frac{1}{3} \cdot \pi \cdot 10^2 \cdot 24 = 800\pi$ cm³

27. $d = 8$ inches, $r = 4$ inches, $b = 3$ inches, $l = 5$ inches, $SA = \pi \cdot 4^2 + \pi \cdot 4 \cdot 5 = 36\pi$ square inches, $V = \frac{1}{3} \cdot \pi \cdot 4^2 \cdot 3 = 16\pi$ cubic inches

28. $C = 16\pi$ cm, $d = 16$, $r = 8$, $b = 6$ cm, $l = 10$ cm, SA $= \pi \cdot 8^2 + \pi \cdot 8 \cdot 10 = 144\pi$ cm², $V = \frac{1}{3} \cdot \pi \cdot 8^2 \cdot 6 = 128\pi$ cm³

29. $r = 12$ inches, $b = 5$ inches, $l = 13$, SA $= \pi \cdot 12^2 + \pi \cdot 12 \cdot 13 = 300\pi$ square inches, $V = \frac{1}{3} \cdot \pi \cdot 12^2 \cdot 5 = 240\pi$ cubic inches

30. $A = 324\pi$ cm, $r = 18$ cm, $b = 24$ cm, $l = 30$ cm, SA $= \pi \cdot 18^2 + \pi \cdot 18 \cdot 30 = 864\pi$ cm², $V = \frac{1}{3} \cdot \pi \cdot 18^2 \cdot 24 = 2,592\pi$ cm³

31. $r = 8$ inches, SA $= 4 \cdot \pi \cdot 8^2 = 256\pi$ square inches

32. $r = 12$ cm, $V = \frac{4}{3} \cdot \pi \cdot 12^3 = 2,304\pi$ cm³

33. $d = 4$ m, $r = 2$m, SA $= 4 \cdot \pi \cdot 2^2 = 16\pi$ m²

34. $d = 6$ feet, $r = 3$ feet, $V = \frac{4}{3} \cdot \pi \cdot 3^3 = 36\pi$ cubic feet

35. $V = 4500\pi$ cm³, $r = 15$ cm

$$\frac{4}{3}\pi r^3 = 4,500\pi$$

$$\frac{4}{3}r^3 = 4,500$$

$$4r^3 = 13,500$$

$$r^3 = 3,375$$

$$r = 15$$

# Chapter 17

1. Shaded area = area of large square – area of 2 white squares $= 7^2 = 2 \cdot 3^2 = 49 - 18 = 31$ square units.

2. Shaded area = area of rectangle – area of two white strips adjusted for overlap of white strips = $20 \times 6 - (20 \times 1 + 6 \times 1 - 1 \times 1) = 120 - 25 = 95$.

3. Shaded area = ½ the area of large circle + ½ the area of small circle $= 72\pi + 12.5\pi = 84.5\pi$.

4. Shaded area $= \frac{1}{2} \times 1 \times 12 = 6$.

5. Flip one shaded section. Shaded area $= \frac{1}{2}(BF + (AH + DE) \times HI = \frac{1}{2}(18 + 16) \times 5 = 85$.

6. $\angle A \cong \angle X$, $\angle B \cong \angle Y$, $\overline{AB} \cong \overline{XY}$, $\triangle ABC \cong \triangle XYZ$ by ASA

7. $\overline{CT} \cong \overline{IN}$, $\angle A \cong \angle W$, $\angle C \cong \angle I$, $\triangle ACT \cong \triangle IWN$ by AAS

8. $\overline{BI} \cong \overline{MA}$, $\overline{IG} \cong \overline{AN}$, $\angle B \cong \angle M$, $\triangle BIG$ and $\triangle MAN$ cannot be determined.

9. $\overline{CA} \cong \overline{DO}$, $\overline{AT} \cong \overline{OG}$, $\overline{CT} \cong \overline{DG}$ $\triangle CAT \cong \triangle DOG$ by SSS

10. $\overline{BO} \cong \overline{CA}$, $\angle O \cong \angle A$, $\overline{OX} \cong \overline{AR}$,
    $\triangle BOX \cong \triangle CAR$ by SAS

11. $\triangle ABC \sim \triangle XYZ$,
    $$\frac{AB}{XY} = \frac{BC}{YZ} = \frac{AC}{XZ}$$

12. $\triangle RST \sim \triangle FED$,
    $$\frac{RS}{FE} = \frac{ST}{ED} = \frac{RT}{FD}$$

13. $\triangle PQR \sim \triangle VXW$,
    $$\frac{PQ}{VX} = \frac{QR}{XW} = \frac{PR}{VW}$$

14. $\triangle MLN \sim \triangle LJK$,
    $$\frac{ML}{LJ} = \frac{LN}{JK} = \frac{MN}{LK}$$

15. $\triangle ZXY \sim \triangle BCA$,
    $$\frac{ZX}{BC} = \frac{XY}{CA} = \frac{ZY}{BA}$$

16. $\triangle GHI \sim \triangle ARM$, $GH = 9$ ft, $GI = 8$ ft, $AR = 12$ ft. Find $AM$.
    $$\frac{9}{12} = \frac{8}{x}$$
    $$9x = 96$$
    $$x = 10\frac{2}{3}$$

17. $\triangle JKL \sim \triangle DOG$, $JK = 17$ m, $JL = 25$ m, $DG = 30$ m. Find $DO$.
    $$\frac{17}{x} = \frac{25}{30}$$
    $$25x = 510$$
    $$x = 20.4$$

18. $\triangle ABC \sim \triangle XYZ$, $AB = 21$ cm, $BC = 54$ cm, $XY = 7$ cm. Find $YZ$.
    $$\frac{21}{7} = \frac{54}{x}$$
    $$21x = 378$$
    $$x = 18$$

19. $\triangle DEF \sim \triangle CAT$, $DE = 65$ in, $EF = 45$ in, $CA = 13$ in. Find $AT$.
    $$\frac{65}{13} = \frac{45}{x}$$
    $$65x = 585$$
    $$x = 9$$

20. $\triangle MNO \sim \triangle LEG$, $MN = x - 3$, $NO = 3$, $EG = 21$, $LE = 2x + 4$. Find $LE$.
    $$\frac{x-3}{2x+4} = \frac{3}{21}$$
    $$21(x-3) = 3(2x+4)$$
    $$21x - 63 = 6x + 12$$
    $$15x - 63 = 12$$
    $$15x = 75$$
    $$x = 5$$

21. $\sin \angle A = \dfrac{BC}{AC} = \dfrac{5}{13}$

22. $\tan \angle BC = \dfrac{AB}{BC} = \dfrac{12}{5}$

23. $\cos \angle A = \dfrac{AB}{AC} = \dfrac{12}{13}$

24. $\tan \angle A = \dfrac{BC}{AB} = \dfrac{5}{12}$

25. $\sin \angle BC = \dfrac{AB}{AC} = \dfrac{12}{13}$

26. $\cos 56° = \dfrac{x}{42}$,
    $x = 42\cos 56° = 23.5$,
    $XY \approx 24$ cm

27. $\tan 46° = \dfrac{x}{42}$,
    $x = 42\tan 46° = 43.5$,
    $BC \approx 44$ cm

28. $\sin 30° = \dfrac{x}{24}$,
    $x = 24\sin 30° = 12$,
    $RS = 12$ cm

29. $\cos 76°30' = \dfrac{80}{x}$,

$x = \dfrac{80}{\cos 76°30'} = 342.7$,

$\cos 76° = \dfrac{80}{x}$,

$x = \dfrac{80}{\cos 76°} = 330.7$,

$AB = 343\ (331)$ feet

30. $\tan 32° = \dfrac{58}{x}$,

$x = \dfrac{58}{\tan 32°} = 92.8$

Side $\overline{YZ}$ is 58 m.
Find the length of $\overline{XY}$ to the nearest centimeter. $XY \approx 93$

# Chapter 18

1. $2 \times 4 = 8$

2. $2 \times 5 \times 4 = 40$

3. $2 \times 5 \times 3 \times 4 = 120$

4. $2 \times 5 \times 4 \times 4 = 160$

5. $2 \times 5 \times 4 \times 3 \times 2 \times 4 = 960$

6. $6! = 6 \times 5 \times 4 \times 3 \times 2 \times 1 = 720$.

7. $7! = 7 \times 6 \times 5 \times 4 \times 3 \times 2 \times 1 = 5{,}040$.

8. $7! \div 6 = 840$.

9. The permutations of 9 things taken 3 at a time $= 504$.

10. The permutations of 10 things taken 4 at a time $= 5{,}040$.

11. The number of combinations of 6 things taken 3 at a time $= 20$.

12. The number of combinations of 7 things taken 4 at a time $= 35$.

13. The number of combinations of 5 things taken 2 at a time $= 10$.

14. The number of different committees of 5 people that can be chosen from a group of 12 people is 792.

15. If you are going to choose 3 toppings from a list of 12 possibilities and the order in which you put toppings on does not matter, you have 220 sundaes to choose from.

16. The probability of drawing a heart and then a queen is $\dfrac{13}{52} \times \dfrac{4}{52} = \dfrac{1}{4} \times \dfrac{4}{52} = \dfrac{1}{52}$.

17. The probability of drawing a heart and then a heart is $\dfrac{13}{52} \times \dfrac{13}{52} = \dfrac{1}{4} \times \dfrac{1}{4} = \dfrac{1}{16}$.

18. The probability of drawing a black card and then a red card is $\dfrac{26}{52} \times \dfrac{26}{52} = \dfrac{1}{2} \times \dfrac{1}{2} = \dfrac{1}{4}$.

19. The probability of drawing a king and a queen is $\dfrac{4}{52} \times \dfrac{4}{51} = \dfrac{16}{2652} = \dfrac{4}{663}$.

20. The probability of drawing two black cards is $\dfrac{26}{52} \times \dfrac{25}{51} = \dfrac{1}{2} \times \dfrac{25}{52} = \dfrac{25}{104}$.

21. The probability that the marble chosen is red or blue is $\dfrac{4+6}{24} = \dfrac{10}{24} = \dfrac{5}{12}$.

22. The probability that the chosen marble is yellow or blue is $\dfrac{2+4}{24} = \dfrac{6}{24} = \dfrac{1}{4}$.

23. The probability that the marble is green or white is $\dfrac{9+3}{24} = \dfrac{12}{24} = \dfrac{1}{2}$.

24. The probability that the chosen marble is yellow or red is $\dfrac{2+6}{24} = \dfrac{8}{24} = \dfrac{1}{3}$.

25. The probability that the marble is red or orange is $\dfrac{6+0}{24} = \dfrac{6}{24} = \dfrac{1}{4}$.

# Chapter 19

1.

**Average Gasoline Usage in Hundreds of Gallons per Month for New England States**

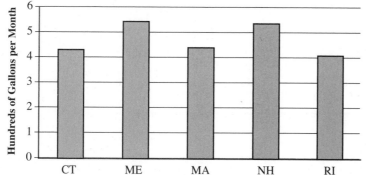

2. Rhode Island had the lowest average gasoline usage, possibly because the small size of the state means commuting distances are smaller.

3. Tacos outsell pasta by approximately 1,000 lunches.

4. Approximately 6,000 chicken lunches are sold per year.

5. Approximately 5,000 burgers are sold per year.

6. Land Area of NYC by Borough

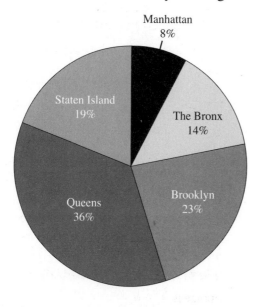

7. Area of Queens ÷ area of Manhattan = $109 \div 23 \approx 4.7$. Queens is between 4 and 5 times the size of Manhattan.

8. Percent of the enrollment in music courses = 4% + 22% + 16% = 42%.

9. Chorus had the largest enrollment.

10. Painting and Ceramics had the most similar enrollments.

11. Art History was 20% of enrollment, so 20% of 461 is approximately 92 students.

12.

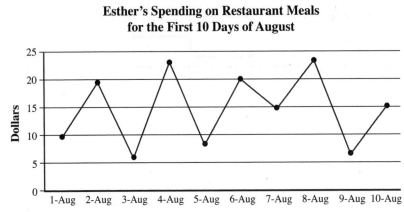

**Esther's Spending on Restaurant Meals for the First 10 Days of August**

13. Esther's restaurant spending alternates large and small expenditures.

14. Esther's spending varied by only 25 cents on August 4th and 8th and on August 7th and 10th.

15. Sales have the greatest positive change from June to July.

16. Sales had the steepest drop from August to September.

17. November had sales most similar to the number of hot dogs sold in February.

# Chapter 20

1. For data set A, mean = $3.\overline{1}$

2. For data set B, mean = 71.8

3. For data set C, mean = 5.3

4. For data set D, mean = 35

5. For data set E, mean = 3.2

6. The median of {2, 2, 2, 3, 3, 4, 4, 4, 4} is 3.

7. The median of {34, 54, 78, 92, 101} is 78.

8. The median of {3, 4, 5, 4, 7, 8, 9, 2, 10, 1} is 4.5.

9. The median of {32, 34, 36, 38} is 35.

10. The median of {2, 2, 3, 4, 5} is 3.

11. The mode of {2, 2, 2, 3, 3, 4, 4, 4, 4} is 4.

12. B = {34, 54, 78, 92, 101} has no mode.

13. The mode of {3, 4, 5, 4, 7, 8, 9, 2, 10, 1} is 4.

14. D = {32, 34, 36, 38} has no mode.

15. The mode of {2, 2, 3, 4, 5} is 2.

16. For data set A, Q1 = 4, Median = 8.5, Q3 = 32.

17. For data set B, Q1 = 3, Median = 4, Q3 = 66.

18. For data set C, Q1 = 25, Median = 64, Q3 = 83.

19. George's placement at the 54th percentile means that his score is slightly above the mean, but Harry's 43rd percentile places him below the mean. George did better.

20.

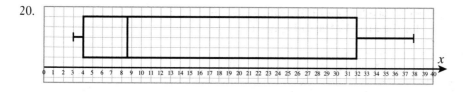

21. The range of {34, 54, 78, 92, 101} is
    $101 - 34 = 67.$

22. The range of {3, 4, 5, 4, 7, 8, 9, 2, 10, 1} is
    $10 - 1 = 9.$

23. The interquartile range of {3, 4, 5, 4, 7, 8, 9,
    2, 10, 17, 32, 34, 36, 38} is $32 - 4 = 28.$

24. The IQR of {2, 2, 2, 3, 3, 4, 4, 4, 4, 34, 54, 78,
    92, 101} is $54 - 3 = 51.$

25. The standard deviation of the test scores
    {69, 70, 73, 74, and 74} is approximately
    2.345.

# Extra Practice Answers

## Part I: Arithmetic

You used the number 8,472,019 to answer questions 1–5.

1. 0 is in the hundreds place.

2. 4 is in the hundred-thousands place.

3. 2 is in the thousands place.

4. 1 is in the tens place.

5. 8 is in the millions place.

6. $7,300 = 7.3 \times 10^3$

7. $12,000 = 1.2 \times 10^4$

8. $903 = 9.03 \times 10^2$

9. $2,450,000 = 2.45 \times 10^6$

10. $691,000 = 6.91 \times 10^5$

11. $8 \times 3 + 5^2 = 24 + 25 = 49$

12. $27 - 8 \div 4 \times 10 = 27 - 2 \times 10 = 27 - 20 = 7$

13. $4(7 - 3) - 3(7 - 4) = 4(4) - 3(3) = 16 - 9 = 7$

14. $(56 - 8) \div 4 - 10 = 48 \div 4 - 10 = 12 - 10 = 2$

15. $8 \times (3 + 5)^2 = 8 \times (8)^2 = 8 \times 64 = 512$

16. $-19 + 39 \div -3 = -19 + -13 = -32$

17. $-5(-4 + -8) + 5(4 + 7) = -5(-12) + 5(11) = 60 + 55 = 115$

18. $(16 - 37) \div -7 = -21 \div -7 = 3$

19. $20 \times -3 + (-7)^2 = 20 \times -3 + 49 = 60 - 49 = 11$

20. $17 - 33 - 9(12 - 8) = 17 - 33 - 9(4) = 17 - 33 - 36 = -16 - 36 = -52$

21. $280 = 2^3 \times 5 \times 7$

22. $5,005 = 5 \times 7 \times 11 \times 13$

23. $510 = 2 \times 3 \times 5 \times 17$ and $272 = 2^4 \times 17$
GCF $= 2 \times 17 = 34$

24. $91 = 7 \times 13$ and $119 = 7 \times 17$
LCM $= 7 \times 13 \times 17 = 1,547$

25. $42 = 2 \times 3 \times 7$ and $90 = 2 \times 3^2 \times 5$
GCF $= 2 \times 3 = 6$ and
LCM $= 2 \times 3^2 \times 5 \times 7 = 630$

26. $\dfrac{15}{28} \times 4\dfrac{2}{3} = \dfrac{\overset{5}{\cancel{15}}}{\underset{2}{\cancel{28}}} \times \dfrac{\cancel{14}}{\cancel{3}} = \dfrac{5}{2} = 2\dfrac{1}{2}$

27. $7\dfrac{3}{5} \div 1\dfrac{9}{10} = \dfrac{38}{5} \div \dfrac{19}{10} = \dfrac{\overset{2}{\cancel{38}}}{\cancel{5}} \times \dfrac{\overset{2}{\cancel{10}}}{\cancel{19}} = \dfrac{4}{1} = 4$

28. $\dfrac{3}{5} + \dfrac{4}{7} = \dfrac{21}{35} + \dfrac{20}{35} = \dfrac{41}{35} = 1\dfrac{6}{35}$

29. $4\dfrac{3}{35} - \dfrac{7}{15} = 4\dfrac{9}{105} - \dfrac{49}{105} = 3\dfrac{114}{105} - \dfrac{49}{105} = 3\dfrac{65}{105} = 3\dfrac{13}{21}$

30. $\dfrac{27}{50} - \dfrac{1}{2} = \dfrac{27}{50} - \dfrac{25}{50} = \dfrac{2}{50} = \dfrac{1}{25}$

31. $3.78 + 9.3 = 13.08$

32. $29.3 - 18.53 = 10.77$

33. $6.23 \times 2.04 = 12.7092$

34. $4.374 \div 1.2 = 3.645$

35. $1.1 \div 0.88 = 1.25$

36. 20.25 ounces of water

$$\dfrac{6}{45} = \dfrac{27}{x}$$
$$6x = 121$$
$$x = 20$$

37. 1.5 pounds of dark chocolate candies

$$\dfrac{3}{3+7} = \dfrac{x}{5}$$
$$10x = 15$$
$$x = 1.5$$

38. 30 cups of wheat flour

$$\dfrac{4}{3} = \dfrac{x}{22.5}$$
$$3x = 90$$
$$x = 30$$

39. 15 gallons of chili

$$\dfrac{6}{40} = \dfrac{x}{100}$$
$$40x = 600$$
$$x = 15$$

40. 66 yards of solid fabric

$$\dfrac{5.5}{7.5} = \dfrac{x}{90}$$
$$7.5x = 495$$
$$x = 66$$

41. 8.1 is 18% of 45.

$$\dfrac{x}{45} = \dfrac{18}{100}$$
$$100x = 810$$
$$x = 8.1$$

42. Thirty-three is 30% of 110.

$$\dfrac{33}{110} = \dfrac{x}{100}$$
$$110x = 3300$$
$$x = 30$$

43. 38 is 16% of 237.5.

$$\dfrac{38}{x} = \dfrac{16}{100}$$
$$16x = 3800$$
$$x = 237.5$$

44. 26% decrease

$$\dfrac{850-629}{850} = \dfrac{x}{100}$$
$$\dfrac{221}{850} = \dfrac{x}{100}$$
$$850x = 22,100$$
$$x = 26$$

45. 12.5% increase

$$\frac{675-600}{600} = \frac{x}{100}$$

$$\frac{75}{600} = \frac{x}{100}$$

$$600x = 7500$$

$$x = 12.5$$

46. $3x - 7x + 12 - 2x + 5 = -6x + 17$

47. $4(5 - 3x) - 2(9x - 7) = 20 - 12x - 18x + 14 = 34 - 30x$

48. $\dfrac{35x - 23 + 15x - 12}{5} = \dfrac{50x - 25}{5} = 10x - 5$

49. $\dfrac{2(5t - 1) + 8t + 2}{9t} = \dfrac{10t - 2 + 8t + 2}{9t} = \dfrac{18t}{9t} = 2$

50. $-7(5y-2) - 8(6-5y) = -35y + 14 - 48 + 40y = y - 34$

51. $7x - 12 = 5x + 4$

    $2x - 12 = 4$

      $2x = 16$

        $x = 8$

52. $3(2x - 9) = -3(x + 5)$

    $6x - 27 = -3x - 15$

    $9x - 27 = -15$

       $9x = 12$

         $x = 1\dfrac{1}{3}$

53. $11t - 49 = 2(7t - 5)$

    $11t - 49 = 14t - 10$

       $-3t = 39$

         $t = -13$

54. $-4y - 6 > -11y + 1$

    $7y - 6 > 1$

      $7y > 7$

       $y > 1$

55. $11 - 3x \le 2x + 1$

    $11 - 5x \le 1$

      $-5x \le -10$

        $x \ge 2$

56. $y = 6 - \dfrac{3}{5}x$

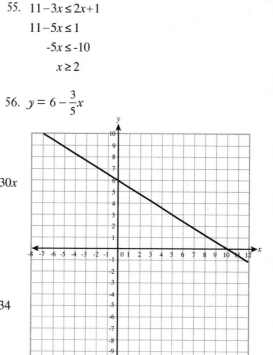

57. $x + y = 8$

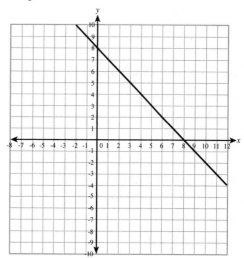

58. $3x - y = 7$

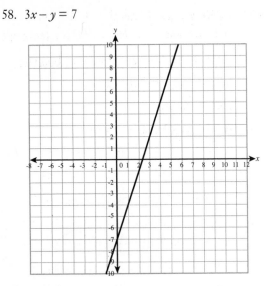

60. $x = -4$

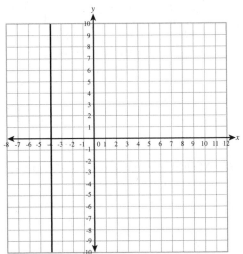

59. $y = 5$

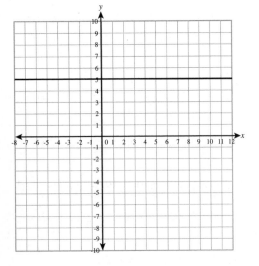

For questions 61 and 62, several answers are possible. Possible answers are given.

61. A ray: $\overrightarrow{RA}$, $\overrightarrow{RP}$, $\overrightarrow{TE}$, and others

62. A line segment: $\overline{IC}$, $\overline{IT}$, $\overline{TC}$, $\overline{PR}$, and others

63. An acute angle: $\angle PRA$

64. An obtuse angle: $\angle ART$

65. A pair of vertical angles: $\angle ITR$ and $\angle CTE$ or $\angle ITE$ and $\angle RTC$

66. The angle that corresponds to $\angle HDE$ is $\angle JEF$.

67. $\angle CDH$ and $\angle BEF$ are a pair of <u>alternate exterior</u> angles.

68. If $m\angle ADC = 20°$, $m\angle DEJ = 180 - 20 = 160°$.

69. The measure of $\angle HJK$ is $90°$.

70. If $m\angle DEJ = 150°$, the measure of $\angle HDE$ is $180 - 150 = 30°$.

71. If $m\angle J = 80°$, $m\angle M\ 80°$, then $m\angle A = 180 - (80° + 80°) = 20°$.

72. If $AR = 7$ cm, $RM = 13$ cm, and $\Delta ARM$ is isosceles, then $AM = \underline{7}$ cm or $\underline{13}$ cm.

73. $\overline{LG}$ must be at least $\underline{13}$ cm but not more than $\underline{21}$ cm.

74. $\Delta MAT$ cannot be a right triangle because $4^2 + 4^2 \neq 6^2$.

75. $\Delta COW$ is a right triangle with right angle $\angle O$. If $CW = 65$ cm and $CO = 25$ cm, then $OW = \sqrt{65^2 - 25^2} = \sqrt{3{,}600} = 60$ cm.

76. If $\Delta PAT \cong \Delta POT$ and $m\angle A = 70°$, then $m\angle O = \underline{70°}$

77. $\Delta DOG \cong \Delta CAT$, and $DO = 28$ inches, $OG = 34$ inches, and $DG = 45$ inches, $CT = DG = 45$ inches

78. $PG = 27$ cm

$$\frac{14}{21} = \frac{18}{x}$$
$$14x = 378$$
$$x = 27$$

79. $RU = 28$ cm

$$\frac{x+8}{x} = \frac{14}{10}$$
$$14x = 10x + 80$$
$$4x = 80$$
$$x = 20$$
$$x + 8 = 28$$

80. $m\angle PAY = m\angle L + m\angle P$, so $115° = 75° + m\angle P$, and $m\angle P = 40°$

81. $ACFH$ is a quadrilateral with $\overline{AC} \parallel \overline{FH}$, $\overline{AH} \parallel \overline{CF}$ and $\overline{CA} \perp \overline{HA}$. $ACFH$ is a <u>rectangle</u>.

82. $BJ = JK = KD = DB$. $BDKJ$ is a <u>rhombus</u>.

83. $\overline{JK} \parallel \overline{GI}$ and $\overline{JI} \parallel \overline{KG}$. $JKGI$ is a <u>parallelogram</u>.

84. In rhombus $BDJK$, $BD = 12$ cm. In parallelogram $JKGI$, $GI = \underline{12}$ cm.

85. $A = bh = 12 \times 5 = 60$ square inches

86. If $m\angle A = 58°$, $m\angle B = 180 - 58 = 122°$.

87. If $BC = 29$ cm and $AD = 41$ cm, $MN = \frac{1}{2}(29 + 41) = \frac{1}{2}(70) = 35$ cm.

88. If $\overline{PQ} \cong \overline{RS}$, the trapezoid is isosceles. If $m\angle Q = 119°$, $m\angle S = m\angle P = 180 - 119 = 61°$.

89. $A = \frac{1}{2}(24 + 18) \times 15 = \frac{1}{2}(42) \times 15 = 21 \times 15 = 315$ cm$^2$

90. The median $= \frac{1}{2}(b_1 + b_2) = 15$.
$A = \frac{1}{2}(b_1 + b_2)h = 15h = 375$, so the height is 25 cm.

91. Arc $\overset{\frown}{AB}$ measures $76°$, so $m\angle APB = \frac{1}{2}(76) = 38°$.

92. $m\angle QPR = \frac{1}{2}(301 - 59) = \frac{1}{2}(242) = 121°$.

93. $m\angle MXS = m\angle NXT = \frac{1}{2}(24 + 52) = \frac{1}{2}(76) = 38°$.

94. $C = 2 \times \pi \times 7 = 14\pi$ cm.

95. Diameter is 24 cm, so $r = 12$ cm.
$A = \pi \times 12^2 = 144\pi$ cm$^2$

96. $SA = 8^2 + 4\left(\frac{1}{2} \cdot 8 \cdot 5\right) = 64 + (4 \cdot 20) = 64 + 80 = 144 \text{cm}^2$

97. $SA = 2\left(\frac{1}{2} \cdot 18 \cdot 24\right) + (18 \cdot 11) + (24 \cdot 11) + (30 \cdot 11) = 432 + 198 + 264 + 330 = 1{,}224 \text{ cm}^2$

98. $SA = 2 \cdot \pi \cdot 15^2 + 2 \cdot \pi \cdot 15 \cdot 22 = 450\pi + 660\pi = 1{,}110\pi$ square inches

99. $SA = \cdot \pi \cdot 5^2 + \pi \cdot 5 \cdot 13 = 25\pi + 65\pi = 90\pi$ square inches

100. $SA = 4 \cdot \pi \cdot 18^2 = 1{,}296\pi \text{ cm}^2$

101. $V = 8 \cdot 5 \cdot 14 = 560$ cubic inches

102. $V = \frac{1}{3} \cdot \left(\frac{1}{2} \cdot 26 \cdot 22.5\right) \cdot 42 = \frac{1}{3} \cdot (292.5) \cdot 42 = 4{,}095$

103. A diameter of 20 cm means a radius of 10 cm. $V = \pi \cdot 10^2 \cdot 48 = 4{,}800\pi \text{ cm}^3$

104. $V = \frac{1}{3} \cdot \pi \cdot 50^2 \cdot 45 = 37{,}500\pi$ cubic inches.

105. A sphere with a diameter of 60 cm has radius of 30 cm. $V = \frac{4}{3} \cdot \pi \cdot 30^3 = \frac{4}{3} \cdot \pi \cdot 27{,}000 = 36{,}000\pi$

106.

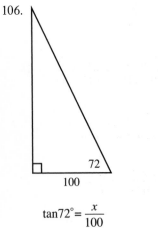

$\tan 72° = \frac{x}{100}$

$x = 100 \tan 72° \approx 307.8$ feet

107.

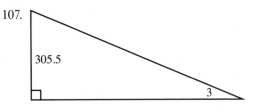

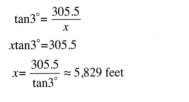

$\tan 3° = \frac{305.5}{x}$

$x \tan 3° = 305.5$

$x = \frac{305.5}{\tan 3°} \approx 5{,}829$ feet

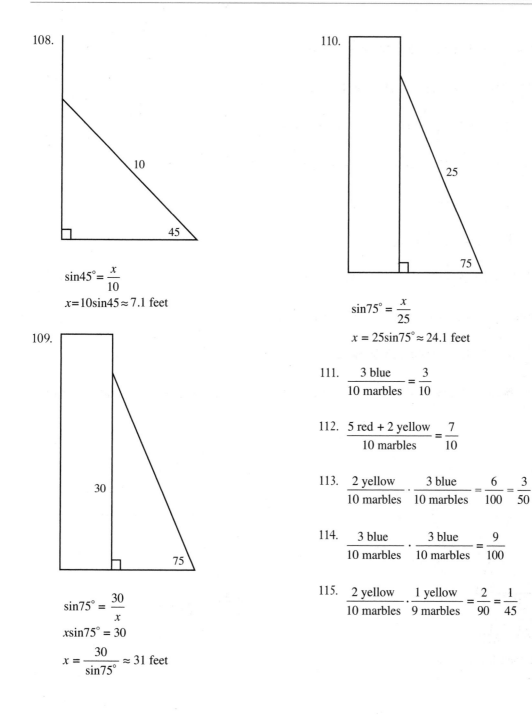

108.

$$\sin45° = \frac{x}{10}$$
$$x = 10\sin45 \approx 7.1 \text{ feet}$$

109.

$$\sin75° = \frac{30}{x}$$
$$x\sin75° = 30$$
$$x = \frac{30}{\sin75°} \approx 31 \text{ feet}$$

110.

$$\sin75° = \frac{x}{25}$$
$$x = 25\sin75° \approx 24.1 \text{ feet}$$

111.  $\dfrac{3 \text{ blue}}{10 \text{ marbles}} = \dfrac{3}{10}$

112.  $\dfrac{5 \text{ red} + 2 \text{ yellow}}{10 \text{ marbles}} = \dfrac{7}{10}$

113.  $\dfrac{2 \text{ yellow}}{10 \text{ marbles}} \cdot \dfrac{3 \text{ blue}}{10 \text{ marbles}} = \dfrac{6}{100} = \dfrac{3}{50}$

114.  $\dfrac{3 \text{ blue}}{10 \text{ marbles}} \cdot \dfrac{3 \text{ blue}}{10 \text{ marbles}} = \dfrac{9}{100}$

115.  $\dfrac{2 \text{ yellow}}{10 \text{ marbles}} \cdot \dfrac{1 \text{ yellow}}{9 \text{ marbles}} = \dfrac{2}{90} = \dfrac{1}{45}$

**Jobs Growth in Selected U.S. Cities**

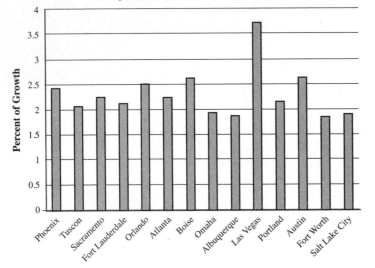

116. Albuquerque and Fort Worth appear to have had the lowest rate of job growth.

117. Three cities—Boise, Las Vegas, and Austin—appear to have job growth rates above 2.5%.

118. According to the graph, the percent of adults smoking cigarettes showed an increase from about 1990 to 1996 with the sharpest increase from 1995 to 1996.

119. According to the graph, the percent of adults smoking cigarettes remained relatively constant from 1985 to 1990.

120.     Favorite Thanksgiving Food

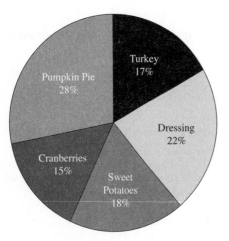

Based on the circle graph, 18% + 15% = 33% of people said their favorite food was a fruit or vegetable.

121. Range = 30 − 6 = 24.

122. Median = 18.

123. First quartile = 14. Third quartile = 26.

124. Mean =
$$\frac{6+12+14+16+16+18+20+22+26+28+30}{11}$$
$$=\frac{208}{11} \approx 18.9$$

125. Standard deviation ≈ 7.23.

| Data | Data – Mean | (Data – Mean)² |
|------|-------------|----------------|
| 6    | -12.9       | 166.64         |
| 12   | -6.9        | 47.74          |
| 14   | -4.9        | 24.10          |
| 16   | -2.9        | 8.46           |
| 18   | -0.9        | 0.83           |
| 20   | 1.1         | 1.19           |
| 22   | 3.1         | 9.55           |
| 26   | 7.1         | 50.28          |
| 28   | 9.1         | 82.64          |
| 30   | 11.1        | 123.01         |

Sum of the squared deviations: 522.91
Divided by 10: ÷ 10 = 52.291
Square root: $\sqrt{52.291} \approx 72.3$

# Glossary

**absolute value**    The distance of a number from zero, without regard to direction.

**acute angle**    An angle that measures less than 90°.

**acute triangle**    A triangle that has three acute angles.

**addend**    The numbers that are added when addition is performed.

**adjacent angles**    A pair of angles that have the same vertex and share a side, but do not overlap one another.

**algorithm**    A list of steps necessary to perform a process.

**alternate interior angles**    Two angles that are on opposite sides of a transversal and between the parallel lines.

**altitude**    A line segment from a vertex of a triangle perpendicular to the opposite side.

**angle**    Two rays with a common endpoint.

**angle bisector**    A line, ray, or segment that passes through the vertex of an angle and cuts it into two angles of equal size.

**angle bisector in a triangle**    A line segment from the vertex of an angle to the opposite side of the triangle that divides the angle into two congruent angles.

**apothem**    A line segment from the center of a polygon perpendicular to a side.

**arc**    A portion of a circle.

**associative property**   A property of addition or multiplication that says that when adding or multiplying more than two numbers you may group them in different ways without changing the result.

**base angles**   The angles at each end of the base, or noncongruent side, of an isosceles triangle.

**binary operation**   A process that works on two numbers at a time.

**binary system**   A place value system based on the number two.

**canceling**   The process of simplifying a multiplication of fractions by dividing a numerator and a denominator by a common factor.

**center**   A point associated with a circle, from which all points of the circle are equidistant.

**central angle**   An angle with its vertex at the center of a circle and sides that are radii.

**chord**   A line segment that connects two points on a circle.

**circle**   The collection of all points that sit at a certain distance, called the radius, from a set point, called the center, forms the shape called a circle.

**circumference**   The distance around a circle.

**circumscribed**   A polygon is circumscribed about a circle if the polygon surrounds the circle with each side tangent to the circle.

**common denominator**   A multiple of the denominators of two or more fractions.

**common fraction**   Fractions written as a quotient of two integers.

**commutative property**   A property of addition or multiplication that says that reversing the order of the two numbers will not change the result.

**compatible numbers**   Two numbers that add to ten.

**complementary angles**   A pair of angles whose measurements total 90°.

**composite number**   A whole number that is not prime because it has factors other than itself and 1.

**concave polygon**   A polygon in which one or more diagonals falls outside the polygon.

**concentric circles**   Circles with the same center.

**cone**   A solid with a circular base and a lateral surface that slopes to a point.

**congruent**   Two segments are congruent if they are the same length. Two angles are congruent if they have the same measure.

**congruent triangles**   Two triangles are congruent if corresponding sides are congruent and corresponding angles are congruent.

**coordinate system**   A system that locates every point in the plane by an ordered pair of numbers, $(x, y)$.

**corresponding angles**   A pair of angles that are on the same side of a transversal and are both above or both below the parallel lines.

**counting numbers**   The set of numbers {1, 2, 3, 4, …} you use to count. The counting numbers are also called the natural numbers.

**cross-multiplying**   Finding the product of the means and the product of the extremes of a proportion and saying that those products are equal.

**cube**   A prism in which all the faces are squares.

**cylinder**   A solid with two circles as parallel bases and a rectangle wrapped around to join them.

**decimal fraction**   Fractions written in the base ten system with digits to the right of the decimal point.

**decimal system**   A place value system in which each position in which a digit can be placed is worth ten times as much as the place to its right.

**denominator**   The number below the division bar in a fraction that tells how many parts the whole was broken into, or what kind of fraction you have.

**diameter**   The longest chord in a circle that passes through the center.

**difference**   The result of a subtraction problem.

**digit**   A single symbol that tells how many.

**distributive property**   The distributive property says that for any three numbers $a$, $b$, and $c$, $c(a + b) = ca + cb$. The answer you get by first adding $a$ and $b$ and then multiplying the sum by $c$ will be the same as the answer you get by multiplying $a$ by $c$ and $b$ by $c$ and then adding the results.

**dividend**   In a division problem, the number that is divided by the divisor.

**divisor**   In a division problem, the number you divide by.

**equation**   A mathematical sentence, which often contains a variable.

**equiangular triangle**   A triangle in which all three angles are 60°.

**equilateral triangle**   A triangle in which all three sides are the same length.

**exponent**    A small number written to the upper right of another number that tells how many of that number should be multiplied together.

**extended ratio**    Several related ratios condensed into one statement. The ratios $a:b$, $b:c$, and $a:c$ make the extended ratio $a:b:c$.

**exterior angle**    The angle formed when one side of a triangle is extended.

**extremes**    The first and last numbers of a proportion.

**faces**    The polygons that connect to form a polyhedron.

**factor**    Each number in a multiplication.

**factor tree**    A method of finding the prime factorization of a number by starting with one factor pair and then factoring each of those factors, continuing until no possible factoring remains.

**fraction**    A symbol that represents part of a whole.

**greatest common factor**    The greatest common factor of two numbers is the largest number that is a factor of both.

**hypotenuse**    In a right triangle, the side opposite the right angle.

**improper fraction**    A fraction whose value is more than one. The numerator is larger than the denominator.

**inscribed**    A polygon is inscribed in a circle if each of its vertices lies on the circle.

**inscribed angle**    An angle with its vertex on the circle and sides that are chords.

**integers**    The set of numbers that includes all the positive whole numbers and their opposites, the negative whole numbers, and zero.

**interest**    Money you pay for the use of money you borrow, or money you receive because you've put your money into a bank account or other investment.

**interest rate**    The percent of the principal that will be paid in interest each year.

**inverse operation**    An operation that reverses the work of another.

**irrational numbers**    Numbers that cannot be written as the quotient of two integers.

**isosceles trapezoid**    A trapezoid in which the nonparallel sides are the same length.

**isosceles triangle**    A triangle with two sides that are the same length.

**lateral area**    The total of the areas of the parallelograms surrounding the bases of a prism. The area of the rectangle that forms the curved surface of a cylinder. The area of the slanted surface of a pyramid or cone.

**least common denominator**   The least common multiple of two or more denominators.

**least common multiple**   The smallest number that has each of two or more numbers as a factor.

**legs**   In a right triangle, the two sides that form the right angle.

**like terms**   Terms that have the same variable, raised to the same power.

**line**   A set of points that has length but no width or height.

**line segment**   A part of a line, made up of two endpoints and all the points of the line between the endpoints.

**linear pair**   Two adjacent angles whose unshared sides form a straight angle.

**major arc**   An arc larger than a semicircle.

**mean**   The arithmetic average of a group of numbers, found by adding all the numbers and dividing by the number of numbers in the group.

**mean of a proportion**   The two middle numbers in a proportion.

**median of a triangle**   A line segment that connects a vertex to the midpoint of the opposite side.

**median of a trapezoid**   The line segment that connects the midpoints of the two non-parallel sides of a trapezoid.

**midpoint**   The point on the segment that divides it into two segments of equal length.

**minor arc**   An arc smaller than a semicircle.

**minuend**   In a subtraction problem, the number from which another number is subtracted.

**mixed number**   A whole number and a fraction, written side by side, representing the whole number plus the fraction.

**natural numbers**   The set of numbers {1, 2, 3, 4, …} you use to count. The natural numbers are also called the counting numbers.

**number line**   A line divided into segments of equal length, labeled with numbers, usually the integers. Positive numbers increase to the right of zero, and negative numbers go down to the left.

**numerator**   The number above the bar in a fraction that tells you how many of that denomination are present.

**obtuse angle**   An angle that measures more than 90° but less than 180°.

**obtuse triangle**   A triangle that contains one obtuse angle.

**ordered pair**    Two numbers, usually designated as $x$ and $y$, that locate a point in a coordinate system.

**order of operations**    An agreement among mathematicians that we perform operations enclosed in parentheses or other grouping symbols first and then evaluate exponents. After that, do multiplication and division as you meet them moving left to right, and finally do addition and subtraction as you meet them, moving left to right.

**parallel lines**    Lines on the same plane that never intersect.

**parallelogram**    A quadrilateral in which both pairs of opposite sides are parallel.

**PEMDAS**    A mnemonic, or memory device, to help you remember that the order of operations is parentheses, exponents, multiplication and division, addition, and subtraction.

**percent**    A ratio that compares numbers to 100. 42% means 42 out of 100, or 42:100.

**perimeter**    The total of the lengths of all the sides of a polygon.

**period**    A group of three digits in a large number. The ones, tens, and hundreds form the ones period. The next three digits are the thousands period, then the millions, the billions, trillions, and on and on.

**permutation**    An arrangement or ordering of a group of objects.

**perpendicular lines**    Lines that meet to form a right angle.

**place value system**    A number system in which the value of a symbol depends on where it is placed in a string of symbols.

**plane**    A flat surface that has length and width but no thickness.

**point**    A position in space that has no length, width, or height.

**polygon**    A closed figure made up of line segments that meet at their endpoints.

**polyhedron**    A solid constructed from polygons that meet at their edges.

**polynomial**    An expression formed by adding terms, each of which is a number times a power of a variable.

**power of ten**    A number formed by multiplying several 10s. The first power of ten is 10. The second power of ten is 100, and the third power of 10 is 1,000.

**prime factorization**    The prime factorization of a number is a multiplication that uses only prime numbers and produces the original number as its product.

**prime number**    A prime number is a whole number whose only factors are itself and 1.

**principal**    The principal is the amount of money borrowed or invested. The rate is the percent of the principal that will be paid in interest each year.

**prism**    A prism is a polyhedron with two parallel faces connected by parallelograms.

**product**    The result of the multiplication is the product.

**proper fraction**    A proper fraction is one whose value is less than one, and an improper fraction is one whose value is more than one.

**proportion**    A proportion is two equal ratios. The means of a proportion are the two middle numbers. The extremes are the first and last numbers.

**protractor**    A protractor is a circle whose circumference is divided into 360 units, called degrees, which is used to measure angles.

**pyramid**    A pyramid is a polyhedron composed of a polygon for a base surrounded by triangles that meet at a point.

**Pythagorean theorem**    A mathematical relation that states that in a right triangle, the square of the hypotenuse is equal to the sum of the squares of the other two sides. It is often expressed as $a^2 + b^2 = c^2$.

**Pythagorean triple**    A Pythagorean triple is a set of three whole numbers $a$, $b$, and $c$ that fit the rule $a^2 + b^2 = c^2$.

**quadrilateral**    A quadrilateral is a polygon with four sides.

**quotient**    The result of a division is called a quotient.

**radius**    The collection of all points that sit at a certain distance, called the radius, from a set point, called the center, forms the shape called a circle.

**rate**    A comparison of two quantities in different units, for example, miles per hour or dollars per day.

**ratio**    A ratio is a comparison of two numbers by division.

**rational numbers**    The set of all numbers that can be written as the quotient of two integers.

**ray**    A ray is a portion of a line from one endpoint, going on forever through another point.

**real numbers**    The name given to the set of all rational numbers and all irrational numbers.

**reciprocal**    Two numbers are reciprocals if their product is 1. Each number is the reciprocal of the other.

**rectangle**    A rectangle is a parallelogram with four right angles.

**regular polygon**     A polygon is regular if all sides are the same length and all angles are congruent.

**relatively prime**     Two numbers are relatively prime if the only factor they have in common is 1.

**remainder**     The number left over at the end of a division problem. It's the difference between the dividend and the product of the divisor and quotient.

**rhombus**     A parallelogram in which all sides are congruent. A square is a parallelogram with four congruent sides and four right angles.

**right angle**     An angle that measures 90°.

**right prism**     A prism is a right prism if the parallelograms meet the bases at right angles.

**right triangle**     A triangle that contains one right angle.

**ruler**     A line or segment divided into sections of equal size, labeled with numbers, called coordinates, used to measure the length of a line segment.

**scale factor**     The scale factor of two similar triangles is the ratio of a pair of corresponding sides.

**scalene triangle**     A triangle that has three sides of different lengths.

**scientific notation**     A method for expressing very large or very small numbers as the product of a number between 1 and 10 and a power of 10.

**secant**     A line that intersects the circle at two different points.

**segment bisector**     A line or ray or segment that passes through the midpoint of a segment and divides a segment into two congruent segments.

**semicircle**     An arc equal to half a circle.

**sides of angle**     The rays with a common endpoint that form the angle.

**similar triangles**     Two triangles are similar if each pair of corresponding angles is congruent and corresponding sides are in proportion.

**slope**     The slope of a line is a number that compares the rise or fall of a line to its horizontal movement.

**solving an equation**     An equation is a mathematical sentence that often contains a variable. Solving an equation is a process of isolating the variable to find the value that can replace the variable to make a true statement.

**space**     The set of all points.

**sphere**     The set of all points in space at a fixed distance from a center point.

**square**   A parallelogram in which all sides are congruent. A square is a parallelogram with four congruent sides and four right angles.

**square numbers**   Numbers created by raising a number to the second power.

**standard form**   When a polynomial is written in standard form, the terms are ordered from highest to lowest degree.

**straight angle**   A straight angle is an angle that measures $180°$.

**subtrahend**   In a subtraction problem, the number subtracted from the minuend.

**sum**   The result of addition.

**supplementary angles**   A pair of angles whose measurements total $180°$.

**surface area**   The total of the areas of all the faces of a solid.

**tangent**   A line that touches a circle at only one point.

**tangent circles**   Circles that touch at only one point.

**term**   An algebraic expression made up of numbers, variables, or both that is connected only by multiplication.

**transversal**   A line that intersects two or more other lines.

**trapezoid**   A quadrilateral in which one pair of opposite sides is parallel.

**unlike terms**   Terms with different variables, such as $x$ and $y$.

**variable**   A letter or symbol that takes the place of a number.

**vertex of an angle**   The point at which the two rays that form the sides of an angle meet.

**vertex of a polygon**   The point at which two sides of a polygon meet.

**vertex angle**   The angle between the equal sides of an isosceles triangle.

**vertical angles**   A pair of angles formed when two lines intersect, which have their vertices at the point where two lines intersect and do not share a side.

**volume**   The volume of a solid is the measure of the space contained by the solid.

**whole numbers**   The set of numbers $\{0, 1, 2, 3, 4, \ldots\}$ formed by adding a zero to the counting numbers.

**$x$-coordinate**   The first number in an ordered pair indicates horizontal movement.

**$y$-coordinate**   The second number in an ordered pair indicates vertical movement.

# Resources

One of the first things you learn as a teacher is that sometimes the best way to help a student understand an idea is to let someone else—not you—explain it. If you need additional guidance on any of the topics in this book, or if you're ready for more math challenges, these resources can help.

For algebra topics that go beyond the scope of this book, check out these *Idiot's Guides*.

Szczepanski, Amy F., PhD, and Andrew P. Kositsky. *The Complete Idiot's Guide to Pre-Algebra*. Indianapolis, IN: Alpha Books, 2008.

Kelley, W. Michael. *The Complete Idiot's Guide to Algebra, Second Edition*. Indianapolis, IN: Alpha Books, 2007.

There are also many, many online math resources. Some are good, some are great, and some are not. When you're looking for math help on the internet, don't immediately believe everything you see. Here are a few resources I recommend.

**Mathforum.org** offers you a chance to have your math questions answered by Dr. Math. Before sending in a question, search the library of questions and answers to see if it has already been answered.

**Coolmath.com** is a colorful site—so colorful it might not seem like a serious math site at first glance—but it's full of tutorials and study tips that you might find helpful.

**Khanacademy.org** has a large collection of video lessons that cover topics from basic arithmetic, through pre-algebra, and on to algebra and beyond.

# Measurement

Throughout this book, you've encountered units of measurement from both the customary and the metric system. This appendix covers the key information you need to function in each system and just a word or two about shifting from one to the other. In each system, you measure three basic quantities: length (or distance), mass (which, with gravity, determines weight), and volume.

## Metric System

The metric system of measurement (also called the International System of Units) is used around the world, and its popularity likely stems not just from the idea of a universal system but from the consistent decimal logic of the system. Everything is based on tens. There are base units of length, mass, and capacity. The larger and smaller units are created by dividing a base unit by 10 or 100 or 1,000 (and so on) or by multiplying by 10 or 100 or 1,000 (and so on).

|  | Basic Unit | Approximation | Official Definition |
|---|---|---|---|
| Length: | meter | The distance from a doorknob to the floor | The path length travelled by light in a given time |
| Mass: | gram* | The mass of a paperclip | The mass of one cubic centimeter of water at 4°C |
| Capacity: | liter | The volume of a medium-sized bottle of soda or water | The capacity of a container with a volume of 1,000 cubic centimeters |

*The current standard defines the kilogram as the base unit and the gram as one one-thousandth of a kilogram, but the system is easier to understand if you begin with gram, the original base unit.

Notice that the units of length, mass, and capacity are linked. The liter is the capacity of a cube 10 centimeters wide by 10 centimeters long by 10 centimeters high, connecting length and volume to capacity. The gram is the mass of a cubic centimeter of water, which associates mass with volume and length and capacity.

From the basic units, you can break into smaller units or build into larger units, always multiplying or dividing by powers of ten. The naming of those units follows the same system of prefixes whether the base unit is meter, liter, or gram. Here are the prefixes and some ideas to help you imagine some of the commonly used units.

| Smaller | | |
|---|---|---|
| $\frac{1}{1,000}$ | milli- | Milli<u>meter</u>: approximately the thickness of 10 sheets of paper<br>Milli<u>gram</u>: the mass of a grain of salt<br>Milli<u>liter</u>: about 20 drops of water |
| $\frac{1}{100}$ | centi- | Centi<u>meter</u>: approximately the diameter of a pencil eraser, or the diameter of a AAA battery<br>Centi<u>gram</u>: the approximate mass of a U.S. dollar bill, or about two raisins.<br>Centi<u>liter</u>: about half a teaspoon |
| $\frac{1}{10}$ | deci- | Deci<u>meter</u>: approximately the length of a crayon<br>Deci<u>gram</u>: two nickels<br>Deci<u>liter</u>: about one-fourth of a can of soda |
| Larger | | |
| 10 | deca-<br>(or deka-) | Deca<u>meter</u>: a long bus or train car<br>Deca<u>gram</u>: about half the mass of a small mouse<br>Deca<u>liter</u>: approximately the capacity of a teapot |
| 100 | hecto- | Hecto<u>meter</u>: about a city block<br>Hecto<u>gram</u>: the mass of an orange<br>Hecto<u>liter</u>: about the capacity of a small refrigerator |
| 1,000 | kilo- | Kilo<u>meter</u>: about 2.5 laps on a stadium track<br>Kilo<u>gram</u>: mass of a dictionary or large textbook<br>Kilo<u>liter</u>: the capacity of about eight large trash cans |

When you're changing units within a system, the thing to remember is balance. If you're changing to a unit of a smaller size, you'll have more of them. If your new unit is bigger, you'll have fewer. And always, it's about 10. Multiply by 10 if you're going to a smaller unit, and divide by 10 to get to a bigger unit.

| | | | Base | | | |
|---|---|---|---|---|---|---|
| 1,000 millimeters | 100 centimeters | 10 decimeters | 1 meter | 0.1 decameter | 0.01 hectometer | 0.001 kilometer |
| 1,000 milligrams | 100 centigrams | 10 decigrams | 1 gram | 0.1 decagram | 0.01 hectogram | 0.001 kilogram |
| 1,000 milliliters | 100 centiliters | 10 deciliters | 1 liter | 0.1 decaliter | 0.01 hectoliter | 0.001 kiloliter |

# Customary System

What's commonly called the customary system is a system that developed over time and remains popular in the U.S. and a few other spots around the world. It is very similar to the British imperial system, as both were derived from English units. The customary system was not designed as a unified system, so there are many different rules to remember.

## Length

In the customary system, length is measured in units like inches, feet, yards, and miles.

| Unit | Approximation | Conversion |
|---|---|---|
| Inch | From the knuckle to the tip of your thumb | |
| Foot | The length of a large man's foot | 12 inches = 1 foot |
| Yard | The height of the kitchen counter | 3 feet = 1 yard |
| Mile | A 20-minute walk | 5,280 feet = 1,760 yards = 1 mile |

## Mass (Weight)

The customary system measures mass (but often calls it weight) in ounces, pounds, and, for really heavy things, tons.

| Unit | Approximation | Conversion |
|---|---|---|
| Ounce | Ten pennies | |
| Pound | A package of butter or bacon, or a football | 16 ounces = 1 pound |
| Ton | A car | 2,000 pounds = 1 ton |

## Capacity

The customary system uses ounce to measure capacity, but an ounce measured by capacity is not necessarily equivalent to an ounce of mass. If you have an ounce of something (capacity), whether it weighs an ounce or not depends on what it is.

| Unit | Approximation | Conversion |
|------|---------------|------------|
| Ounce | A little container of coffee cream | |
| Cup | A container of coffee | 8 ounces = 1 cup |
| Pint | A small container of ice cream | 2 cups = 1 pint |
| Quart | A container of milk | 2 pints = 1 quart |
| Gallon | A large can of paint | 4 quarts = 1 gallon |

# Conversion

Although you may not often need to convert from one system of measurement to another, when you are, you may not have the tools or formulas handy to make an exact conversion. Here are some rules of thumb that can help you make an approximate conversion.

- An inch is about 2.5 centimeters.

- A meter is a little more than a yard. A meter is about 39 inches and a yard is 36 inches.

- A mile is about 1.6 kilometers.

- A liter is a little more than a quart.

- A kilogram is about 2.2 pounds.

> **WORLDLY WISDOM**
>
> A famous number sequence, called the Fibonacci sequence, is easy to re-create and can be used for quick, approximate conversions of length and distance. The Fibonacci sequence begins with two ones, then forms each of the next terms by adding the two previous terms.
>
> 1, 1, 1 + 1
>
> 1, 1, 2, 1 + 2
>
> 1, 1, 2, 3, 2 + 3
>
> 1, 1, 2, 3, 5, 8, 13, 21,...
>
> Two adjacent terms of the Fibonacci sequence can give you a rough conversion of miles and kilometers. For example, 5 kilometers is approximately 3 miles, (3.10686 miles) and 8 kilometers is approximately 5 miles (4.97097).

# Index

## Numbers

## A

## B